# Lecture Notes in Business Information Processing     **340**

More information about this series at http://www.springer.com/series/7911

Paolo Ceravolo · Maurice van Keulen
Kilian Stoffel (Eds.)

# Data-Driven Process Discovery and Analysis

7th IFIP WG 2.6 International Symposium, SIMPDA 2017
Neuchatel, Switzerland, December 6–8, 2017
Revised Selected Papers

 Springer

*Editors*
Paolo Ceravolo ⓘ
Università degli Studi di Milano
Crema, Italy

Maurice van Keulen ⓘ
University of Twente
Enschede, The Netherlands

Kilian Stoffel
IMI
University of Neuchâtel
Neuchâtel, Switzerland

ISSN 1865-1348 ISSN 1865-1356 (electronic)
Lecture Notes in Business Information Processing
ISBN 978-3-030-11637-8 ISBN 978-3-030-11638-5 (eBook)
https://doi.org/10.1007/978-3-030-11638-5

Library of Congress Control Number: 2018967449

This Springer imprint is published by the registered company Springer Nature Switzerland AG
The registered company address is: Gewerbestrasse 11, 6330 Cham, Switzerland

# Preface

The rapid growth of organizational and business processes data, managed via information systems, has made available a big variety of information that consequently created a high demand for making data analytics more effective and valuable. The seventh edition of the International Symposium on Data-Driven Process Discovery and Analysis (SIMPDA 2017) was conceived to offer a forum where researchers from different communities can share their insights in this hot new field. As a symposium, SIMPDA fosters exchanges among academic research, industry, and a wider audience interested in process discovery and analysis. The event is organized by the IFIP WG 2.6. This year the symposium was held in Neuchatel, Switzerland.

Submissions cover theoretical issues related to process representation, discovery, and analysis or provide practical and operational examples of their application. To improve the quality of the contributions, the symposium is structured for fostering discussion and stimulating improvements. Papers are pre-circulated to the authors, who are expected to read them and make ready comments and suggestions. After the event, authors have the opportunity to improve their work extending the presented results. For this reason, authors of accepted papers were invited to submit extended articles to this post-symposium volume. We received 19 submissions and six papers were accepted for publication in this volume.

The current selection of papers underlines the most relevant challenges that were identified and proposes novel solutions for facing these challenges.

In the first paper, "Online Detection of Operator Errors in Cloud Computing Using Anti-Patterns," Arthur Vetter studies the role of anti-patterns to support monotonic inference in real-time event processing. In particular his word addresses monitoring on a specific model, namely, the topology and orchestration specification for cloud applications, which explicitly models the maintenance operations of IT service applications.

The second paper, by Sebastian Steinau et al., is titled "Executing Lifecycle Processes in Object-Aware Process Management" and presents an advanced methodology for coping with object-aware process management, where the operational semantics is not obtained by specifying a workflow but by constraining the data flow characterizing business objects.

The third paper by Leonardi et al., "Towards Semantic Process Mining Through Knowledge-Based Trace Abstraction" proposes an approach to lift the semantics of event logs. The proposed framework is able to convert actions found in the event log into higher-level concepts, on the basis of a domain knowledge. According to the authors, the semantics lift process is proven to be a means to significantly increase the quality of the mined models, when measured in terms of fitness.

The fourth paper by Gega et al., "Mining Local Process Models and Their Correlations" aims at simplifying the integration of local process model (LPM) mining, episode mining, and the mining of frequent subtraces. For instance, the output of a subtrace mining approach can be used to mine LPMs more efficiently. Also, instances of LPMs can be correlated together to obtain larger LPMs, thus providing a more comprehensive overview of the overall process. The authors discuss the benefit of this integration on a collection of real-life event logs.

The fifth paper by Couvreur and Ezpeleta, "A Linear Temporal Logic Model-Checking Method over Finite Words with Correlated Transition Attributes" presents an adaption of the classic timed propositional temporal logic to the case of finite words and considers relations among different attributes corresponding to different events. The introduced approach allows for the use of general relations between event attributes by means of freeze quantifiers as well as future and past temporal operators. The paper also presents a decision procedure, as well as a study of its computational complexity.

The sixth paper by Azzini et al., "A Report-Driven Approach to Design Multidimensional Models" presents an approach that can generate a multidimensional model from the structure of expected reports as data warehouse output. The approach is able to generate the multidimensional model and populate the data warehouse by defining a knowledge base specific to the domain. Although the use of semantic information in data storage is not new, the novel contribution of this approach is represented by the idea of simplifying the design phase of the data warehouse, making it more efficient, by using an industry-specific knowledge base and a report-based approach.

We gratefully acknowledge the research community that gathered around the problems related to process data analysis. We would also like to express our deep appreciation of the referees' hard work and dedication. Above all, thanks are due to the authors for submitting the best results of their work to the Symposium on Data-Driven Process Discovery and Analysis.

We are very grateful to the Università degli Studi di Milano and to IFIP for their financial support, and to the University of Neuchatel for hosting the event.

November 2018                                              Paolo Ceravolo
                                                     Maurice Van Keulen
                                                           Kilan Stoffel

# Organization

## Chairs

| | |
|---|---|
| Paolo Ceravolo | Università degli Studi di Milano, Italy |
| Maurice Van Keulen | University of Twente, The Netherlands |
| Kilan Stoffel | University of Neuchatel, Switzerland |

## Advisory Board

| | |
|---|---|
| Ernesto Damiani | Università degli Studi di Milano, Italy |
| Erich Neuhold | University of Vienna, Austria |
| Philippe Cudré-Mauroux | University of Fribourg, Switzerland |
| Robert Meersman | Graz University of Technology, Austria |
| Wilfried Grossmann | University of Vienna, Austria |

## SIMPDA Award Committee

| | |
|---|---|
| Paul Cotofrei | University of Neuchatel, Switzerland |
| Paolo Ceravolo | Università degli Studi di Milano, Italy |

## Web and Publicity Chair

| | |
|---|---|
| Fulvio Frati | Università degli Studi di Milano, Italy |

## Program Committee

| | |
|---|---|
| Akhil Kumar | Penn State University, USA |
| Benoit Depaire | University of Hasselt, Belgium |
| Chintan Amrit | University of Twente, The Netherlands |
| Christophe Debruyne | Trinity College Dublin, Ireland |
| Ebrahim Bagheri | Ryerson University, Canada |
| Edgar Weippl | TU Vienna, Austria |
| Fabrizio Maria Maggi | University of Tartu, Estonia |
| George Spanoudakis | City University London, UK |
| Haris Mouratidis | University of Brighton, UK |
| Isabella Seeber | University of Innsbruck, Austria |
| Jan Mendling | Vienna University of Economics and Business, Austria |
| Josep Carmona | UPC - Barcelona, Spain |
| Kristof Boehmer | University of Vienna, Austria |
| Manfred Reichert | Ulm University, Germany |
| Marcello Leida | TAIGER, Spain |
| Mark Strembeck | WU Vienna, Austria |

| | |
|---|---|
| Massimiliano De Leoni | Eindhoven TU, Netherlands |
| Matthias Weidlich | HU Berlin, Germany |
| Mazak Alexandra | Vienna University of Technology, Austria |
| Mohamed Mosbah | University of Bordeaux, France |
| Mustafa Jarrar | Birzeit University, Palestine |
| Robert Singer | FH JOANNEUM, Austria |
| Roland Rieke | Fraunhofer SIT, Germany |
| Schahram Dustdar | Vienna University of Technology, Austria |
| Thomas Vogelgesang | University of Oldenburg, Germany |
| Valentina Emilia Balas | University of Arad, Romania |
| Wil Van der Aalst | Technische Universiteit Eindhoven, The Netherlands |

# Contents

# Online Detection of Operator Errors
# in Cloud Computing Using Anti-patterns

Arthur Vetter[✉]

Horus software GmbH, Ettlingen, Germany
arthur.vetter@horus.biz

**Abstract.** IT services are subject of several maintenance operations like upgrades, reconfigurations or redeployments. Monitoring those changes is crucial to detect operator errors, which are a main source of service failures. Another challenge, which exacerbates operator errors is the increasing frequency of changes, e.g. because of continuous deployments like often performed in cloud computing. In this paper, we propose a monitoring approach to detect operator errors online in real-time by using complex event processing and anti-patterns. The basis of the monitoring approach is a novel business process modelling method, combining TOSCA and Petri nets. This model is used to derive pattern instances, which are input for a complex event processing engine in order to analyze them against the generated events of the monitored applications.

**Keywords:** Complex event processing · Anti-pattern · TOSCA ·
IT service management · Anomaly detection

## 1 Introduction

Operator errors have been one of the major reasons for IT service failures [1–6] and will probably continue to be regarding current trends like continuous delivery, DevOps and infrastructure-as-code [7], which are especially very common in cloud computing. In recent years, several studies and methods were developed to detect errors in very complex IT systems [8]. Those traditional methods are suited for detecting errors during "normal" operations, but not during change operations like reconfigurations or rolling upgrades, when one node after the other is upgraded [9]. The reason for the amount of operator errors is their human nature, because change operations are either performed or initiated by human operators.

This paper presents current research results of a novel monitoring approach for those change operations. The monitoring approach is based on a process model, combining TOSCA (Topology and Orchestration Specification for Cloud Applications) and high-level Petri nets [8], which explicitly models the maintenance operations of the IT service applications. This process model is used to derive pattern instances from it. Those pattern instances are checked through

© IFIP International Federation for Information Processing 2019
Published by Springer Nature Switzerland AG 2019
P. Ceravolo et al. (Eds.): SIMPDA 2017, LNBIP 340, pp. 1–24, 2019.
https://doi.org/10.1007/978-3-030-11638-5_1

a complex event processing engine against state events and transaction events. State events describe the state of the application, whereas transaction events describe each single operation performed on the application. Therefore, the logs of the applications are filtered for meaningful transaction events and are sent to the complex event processing engine, allowing the detection of operator errors almost in real-time. The complex event processing engine compares the pattern instances with the generated events through anti-patterns and creates an error message, when an anti-pattern instance was detected. Figure 1 gives an overview of the general monitoring approach.

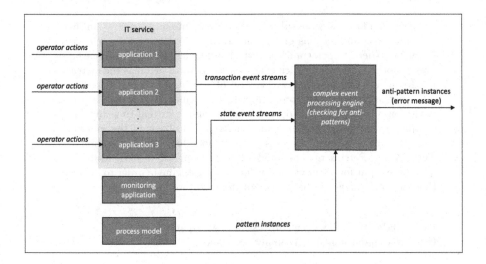

**Fig. 1.** General monitoring approach

The remainder of this paper is organized as follows: The next section gives a short overview of typical operator errors. Section 3 describes the fundamentals of TOSCA and XML nets, which are used to model the actual maintenance. Section 4 describes the concept of patterns and anti-patterns. Section 5 presents the proof of concept implementation. In the next section first experimental results are presented and discussed in the following section. Afterwards related work is presented. Section 9 concludes the paper.

## 2   Operator Errors

Oppenheimer et al. [5] and many other authors like [4,9] classify operator errors in process errors and configuration errors. Process errors can be further differentiated in following errors: forgotten activity, an unneeded activity was executed, a wrong activity was executed or actual correct activities were executed in the wrong order. Configuration errors can be separated in formatting errors and configuration value errors [13]. Formatting errors can be further separated in lexical errors, syntactical errors and typos. Configuration value errors can be further classified in local value inconsistencies and global environment inconsistencies. A monitoring approach to detect operator errors should be able to detect all those process and configuration error types. Table 1 gives an example for every type of operator error and a reference to a study with further information and examples.

**Table 1.** Operator error examples

| Operator error | Example | Description | Reference |
|---|---|---|---|
| Forgotten activity | Forgot to restart a server | | [4] |
| Unneeded activity | Unnecessary restart of a server | | [9] |
| Wrongly executed activity | Restoration of a wrong backup | | [4] |
| Wrong order | Bringing down two servers in parallel for configuration instead of sequentially maintaining the servers | | [9] |
| Local Inconsistency | log_output = "Table" log = query.log | According to the value "log", the user wanted to store logs in a file, but the value "log.output" controls to store data in a database table | [10] |
| Global Inconsistency | datadir = /some/old/path | "datadir" points to an old path, which does not exist anymore. | [10] |
| Lexical Errors | InitiatorName: iqn:DEV_domain | Only lowercase letters are allowed ("DEV") | [10] |
| Syntactical Errors | extension = mysql.so ..... extension = recode.so | "mysql.so" depends on "recode.so" and was configured in the wrong order | [10] |
| Typo | extension = recdoe.so extension = mysql.so | The correct writing of "recdoe.so" is "recode.so" | [10] |

# 3   Fundamentals

The process model is a combination of TOSCA and XML nets and was introduced in a former paper [8]. In this chapter, we describe the fundamentals of TOSCA and XML nets shortly and then describe how maintenance operations can be modelled with TOSCA and XML nets.

## 3.1   TOSCA

TOSCA is a standard, released by OASIS [14] to support the portability of cloud applications between different cloud providers and the automation of cloud application provisioning. Therefore, TOSCA provides a modelling language to describe cloud applications as Service Templates. A Service Template consists of a Topology Template and of optional Plans (see Fig. 2).

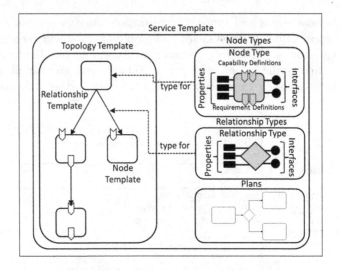

**Fig. 2.** TOSCA service template

A Topology Template describes the structure of a cloud application as a directed graph and consists of Node Templates and Relationship Templates. A Node Template represents a component of the cloud application, e.g. an application server and is described by a Node Type. A Node Type defines

- properties of the component (*Properties Definition*),
- available operations to manipulate the component (*Interfaces*),
- requirements of the component (*Requirement Definitions*),
- possible lifecycle states of the component (*Instance States*) and
- capabilities it offers to satisfy other components' requirements (*Capability Definitions*).

Plans are models to orchestrate the management Operations, which are offered by the cloud application components and can be written in BPMN, BPEL or other languages.

As TOSCA Service Templates are written as XML documents, we decided to use the notation of XML nets for the creation of Plans, which we name "maintenance plan" in the rest of the paper. Using XML nets has the advantage that no additional notation elements have to be defined like it is the case e.g. for BPMN [25]. Apart of that, XML nets allow to describe detailed manipulations of XML documents, which are used to model configuration operations in maintenance plans.

### 3.2   XML Nets

XML nets [15] are a high-level variant of Petri nets, in which places represent containers for XML documents. The XML documents must conform to the XML

Schema, which is assigned to a specific place. Edges are labeled with Filter Schemas, which are used to read or manipulate XML documents. Transitions can be inscribed by a logical expression, whose variables are contained in the adjacent edges. A transition in an XML net is enabled and can be fired for a given marking, when the following three conditions hold. First, every place in the pre-set of the transition holds at least one valid XML document, which conforms to the Filter Schema inscribing the edge to the transition. Second, every place in the post-set of a transition must contain one valid XML document, if the XML document has to be modified. If an XML document has to be created from scratch the place must not already contain this XML document. Third, for the given instantiation of the variables, the transition inscription has to be evaluated to true in order to enable the transition. If an enabled transition is fired, XML documents in the pre-set places are (partially) deleted or read for the given instantiation of variables, and new XML documents are created or existing XML documents are modified in the post-set places of the transition.

### 3.3 Modelling Maintenance Plans

This section describes the modelling of maintenance plans with TOSCA and XML nets, which allows to model applications and the orchestration of applications' management operations in one integrated model. Such a model can then be used to derive pattern instances. Therefore, we extend our former approach, introduced in [8]. The following adjustments are made to the general definition of TOSCA Node Templates:

- A Node Template represents exactly one instance of an application, that means the attributes *minIstances, maxInstances* $:= 1$.
- Node Templates are extended with the complex element *InstanceState*, which stores the current state of the corresponding application.

The notation of XML nets is adjusted as follows:

- Places are containers for *Service Templates*. Every place is assigned to the general TOSCA XML schema and additionally to a single *Node Type*, which restricts the allowed filter schemas for corresponding *Node Templates*.
- Transitions represent operations, defined in *Interfaces* of the adjacent *Node Types*.
- Filter Schemas can either be used to select *Node Templates* or to modify *Properties*, or *Instance States* of a *Node Template*. Deleting whole *Node Templates* is in contrast to general XML nets not allowed. *Node Templates* can only change their status, e.g. to undeploy, but they cannot be deleted. The reason is, that for error detection purposes, even an undeployed Node has to be monitored to be sure it was really undeployed and e.g. has not been deployed by accident afterwards again. Deleting parts of a *Node Template* are allowed, e.g. deleting a property.
- Transitions hold the attributes *start* and *end*, which define when the operation has to be executed earliest and latest.

We define a maintenance plan as a tuple MP = $<P, T, A, \Psi, I_P, I_N, I_A, I_T, M_0>$, where

(i) $<P, T, A>$ is a Petri net with a set of places P, a set T of transitions, and a set A of edges connecting places and transitions (the definition and description of petri nets is excluded in this paper, but can be found, e.g., in [11]).

(ii) $\Psi = <D, FT, PR>$ is a structure consisting of a finite and non-empty individual set D, a set of term and formula functions $FT$ defined on $D$, and a set of predicates $PR$ defined on $D$.

(iii) $I_P$ is the function that assigns the TOSCA XML Schema to each place.

(iv) $I_N$ is the function that assigns additionally a Node Type to each place.

(v) $I_A$ is the function that assigns a Filter Schema to each edge. The Filter Schema must conform to the XML Schema and Node Type of the adjacent place.

(vi) $I_T$ is the function that assigns a predicate logical expression as inscription to each transition. The inscription is built on a given structure $\Psi$ and a set of variables. Only variables, which are contained in the Filter schemas of adjacent arcs, are allowed. The inscription must evaluate to true in order to enable the transition.

(vii) Each transition represents a value of the element *operation*, which is defined in the complex element *Interfaces* of the Node Type in the postset of the transition.

(viii) $M_0$ is the initial marking. Markings are TOSCA Service Templates.

(ix) Each transition holds the attributes *start* and *end*.

Figure 3 shows an example of a maintenance plan to configure the database connection of the application *MyApplication* (Filter Schemas are written informally for readability reasons). It is assumed that the database and application are part of the Service Template *MyService*. *MyApplication* is hosted on *MyAppServer* and requires additionally the database *TestDatabase* (NT1). It is assumed, that when the change is performed, *MyApplication* is started. In the first place, which is linked to a Node Type *Application*, *MyApplication* is one possible representation. The first Filter Schema *FS1* selects *MyApplication*. Before *MyApplication* can be configured it has to be stopped, which is represented in the first transition. The condition in order to stop the application is, that *MyApplication* has to be started. Stopping is one possible operation, which is given by the Node Type *Application*. If at the beginning of executing the change, *MyApplication* is already stopped instead of started, it is a hint, that an incident or something unexpected happened, so the change execution should be interrupted. When *MyApplication* is stopped, the database connection can be set. Therefore, the Node Template *TestDatabase* is selected and the database connection is built up on the properties of *TestDatabase* and inserted in *MyApplication* through the Filter Schema *FS5*. Afterwards *MyApplication* can be started again, but only if *TestDatabase* is running (inscription assigned to transition *Start*).

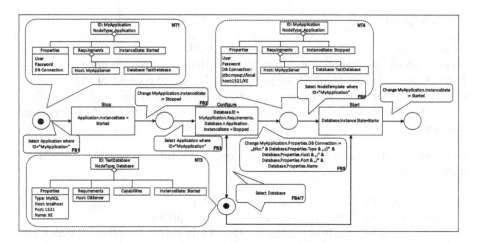

**Fig. 3.** Example of a TOSCA based XML net

# 4 Pattern and Anti-pattern for Operator Error Detection

In computer science the term pattern is popular since the publication of the book about design patterns from Gamma et al. [12]. In this book, Gamma et al. describe patterns as solutions for recurring problems in a specific context. Aalst et al. [13] used the concept of patterns for business process modelling and described several patterns for the control flow perspective. Since then, many patterns were described for different perspectives of business process modelling, like for the data perspective [14,15]. Riehle and Züllighoven define a pattern more general as an abstraction of a recurring concrete form in a specific context [16]. A form is a finite number of distinguishable elements and their relationships [16]. A context restricts the possible usage of a form, because the form has to fit into this specific context. Based on the definition of Riehle and Züllighoven we define a pattern and anti-pattern as following:

**Definition 4.1 (Pattern).** A pattern is an abstraction of a welcomed, recurring, concrete form in a specific context.

**Definition 4.2 (Anti-pattern).** An anti-pattern is as an abstraction of an unwelcomed, concrete form in a specific context.

In our work, we use patterns to describe the planned to be control flow, application configurations and application states for the scheduled maintenance. So, patterns are used during the design phase. Anti-patterns are used to check during the actual execution of the maintenance (run-time), if a form of events exists, which does not fit to the planned forms. In the following we restrict and formalize the context of the used patterns and anti-patterns as well as the form of these patterns and anti-patterns.

## 4.1 Context

As described in Sect. 3, the monitoring approach is based on the comparison between produced events of monitored applications and pattern instances of the TOSCA management plan. Those parameters build the context of the patterns. We separate two kinds of events in our context: *state events* and *transaction events*. Definitions 4.3 and 4.4 formalize *state events* and *transaction events* in this paper.

**Definition 4.3 (State Event).** A state event is a tuple se = (timestamp, app, state), where:

- *timestamp* is the timestamp of the event creation.
- *app* is the *Node Template id* of the monitored application.
- *state* is the actual state of the application. Only values are allowed, which are defined in the Node Type of the application by the element *Instance States*.

The set of all state events is defined as *SES*.

**Definition 4.4 (Transaction Event).** A transaction event is a tuple $te =$ *(timestamp, st, app, op, prop, value)*, where:

- *timestamp* is the timestamp of the event creation.
- *st* is the *Service Template id*, which identifies the service the application belongs to.
- *app* is the Node Template id of the monitored application.
- *op* describes the operation, which was conducted on the application. The value of *op* must correspond to one of the values, which are defined in the element *operation* of the *Node Type* of the application.
- *prop* describes the property, which was changed when the operation was executed. If no property was changed during the operation *prop* is null.
- *value* is the value of the property, which was changed. If *prop* is null, *value* also has to be null.

The set of all transaction events is defined as *TES*. State events and transaction events represent the actual events during a maintenance. The corresponding "to-be" events are conditions and activities, which can be derived from a TOSCA management plan. A condition represents a possible transition inscription, whereas activities represent firing sequences.

**Definition 4.5 (Condition).** A condition is a tuple *(app, op, prop, zapp, state)*, where:

- *app* is the id of the Node Template, on which the operation is performed.
- *op* is the operation, which is performed on the Node Template and is restricted in the Node Type of the Node Template.
- *prop* is the property of the Node Template, which is changed during the operation.

- *zapp* is the id of the Node Template, which has to be in a specific state in order to perform the operation.
- *state* describes in which state *zapp* has to be.

Let *SM* be the set of all maintenance plans. The set of all conditions of a maintenance plan is defined as $SC_i$, $i \in SM$. The set of all transition inscriptions of a maintenance plan is defined as $STI_i$ $i \in SM$. The function $F: SC_i \rightarrow STI_i$ assigns a transition to each condition.

**Definition 4.6 (Activity).** An activity is a tuple $a = $ *(st, app, op, prop, value, start, end)*, where:

- *st* is the *Service Template id*, which identifies the service template in the TOSCA management plan.
- *app* is the id of the Node Template, on which the operation is performed.
- *op* is the operation, which is performed on the Node Template and is restricted in the Node Type of the Node Template.
- *prop* is the property of the Node Template, which is changed during the operation.
- *value* is the value of the property, which was changes. If prop is null, value also has to be null.
- *start* describes when the activity has to start earliest.
- *end* describes when the activity has to end latest.

Be *SM* the set of all maintenance plans. The set of all activities of a maintenance plan is defined as $SA_i$, $i \in SM$. The set of all transitions of a maintenance plan is defined as $ST_i$, $\in SM$. The function $F: SA_i \rightarrow ST_i$ assigns a transition to each activity.

Additionally, for some anti-patterns we need the history of transaction events and the latest state of an application called the state event history.

**Definition 4.7 (Transaction Event History).** A *transaction event history* is a selection on the set of transaction events, which are in the time scope of the scheduled maintenance:

$$TEH := \sigma_{\text{timestamp} \geq \text{maintenance\_start} \, \wedge \, \text{timestamp} \leq \text{maintenance\_end}} TES$$

**Definition 4.8 (State Event History).** The *state event history* SEH stores the latest state for each application in *SES*.

Furthermore, we define three functions, time, countTE and countA.

**Definition 4.9 (Time).** time is a function, which returns the current timestamp.

**Definition 4.10 (CountTE).** *countTE(te, TEH)* is a function, which counts the number of occurrences of the transaction event *te* in the transaction event history.

**Definition 4.11 (CountA).** *countA(a, S)* is a function, which counts the number of occurrences of an activity *a* in a set S.

After the description and definition of the context, the patterns and anti-patterns are described.

## 4.2  Pattern and Anti-pattern

All in all, we define ten patterns/anti-patterns in order to detect operation errors. These are NEXT, IMMEDIATELY NEXT, PRECEDENCE, IMMEDI-ATELY PRECEDENCE, OCCURRENCE, ALTERNATIVE OCCURRENCE, ABSENCE, ALTERNATIVE ABSENCE, VALUE and STATE-CONDITION. The first eight patterns are highly influenced by the specification pattern of Dwyer et al. [17] and are used to detect process errors. Whereas the VALUE anti-pattern is used to detect configuration errors. The STATE-CONDITION anti-pattern is used to check, if a resource is in the planned state in order to perform a task on it. To describe the patterns and anti-patterns following template is used:

- **Name:** The name of the pattern must be unique and should describe the purpose of the pattern.
- **Description:** Here the form of the pattern is described, which should occur in the maintenance.
- **Instances:** Here it is described, how instances of the pattern can be derived from the maintenance plan.
- **Example:** Here, examples of pattern instances are given.
- **Anti-pattern:** A description of the corresponding anti-pattern and which type of operator errors can be detected with the anti-pattern. Additionally, we formalize the conditions, which have to be violated in order to detect an operator error.
- **Similar pattern:** Here, similar patterns are referenced and differences are named.

**Pattern NEXT**
**Description:** This pattern describes pairs of activities, defining which activity has to occur after another (with possible activities inbetween). The pattern is used for controlling AND-joins, AND-splits and concurrent sequences in a maintenance plan.
**Instances:** To get all instances of this pattern for a TOSCA management plan i we create a relation $P1 := AM_i \times AM_i \times AM_i$ with the tuples $(a_{cur}, a_{nex}, a_{far})$ where,

- the corresponding transitions $t_{cur}$ and $t_{nex}$ of the activities $a_{cur}$ and $a_{nex}$ are connected through the same place,
- $t_{cur}$, $t_{nex}$ and $t_{far}$ have to occur in the same path,
- $t_{far}$ always has to occur after $t_{cur}$,
- $t_{far}$ and $t_{cur}$ may not be connected through the same place.

**Example:** In Fig. 4 instances of the pattern NEXT are (a1, a2, a4), (a1, a2, a6), (a1, a3, a5), (a1, a3, a7), (a2, a4, a6) and (a3, a5, a7).

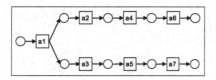

**Fig. 4.** Example pattern NEXT

**Anti-pattern:** The anti-pattern allows to detect operator errors of the type "wrong order". Besides, it is possible to detect operator errors of the type syntactical error, if a configuration parameter was changed in the wrong order. An error message is created, when a transaction event $te_{cur}$ in the event stream occurs and none of the next events $te_{nex}$ conforms to the next activity $a_{nex}$. However, one of the next events conforms to an activity $a_{far}$:

$$\pi_{app,op,prop} te_{cur} \in \pi_{a_{cur}.app,a_{cur}.op,a_{cur}.prop} P1_i \succ \pi_{app,op,prop} te_{nex} \in$$
$$\pi_{app,op,prop} \left( \pi_{a_{far}} (\sigma_{a_{cur}.app=te_{cur}.app \wedge a_{cur}.op=te_{cur}.op \wedge a_{cur}.prop=_{cur}.prop} P1_i) \right) \wedge$$
$$\pi_{app,op,prop} \left( \pi_{a_{nex}} \right($$
$$\sigma_{a_{cur}.app=te_{cur}.app \wedge cur.op=te_{cur}.op \wedge a_{cur}.prop=te_{cur}.prop} \wedge$$
$$a_{far}.app=te_{nex}.app \wedge a_{far}.op=te_{nex}.op \wedge a_{far}.prop=te_{nex}.prop} P1_i) \notin$$
$$\pi_{app,op,prop} (\sigma_{te_{cur}.timestamp > timestamp \wedge te_{nex}.timestamp < timestamp} TEH)$$

**Similar Pattern:** The pattern IMMEDIATELY NEXT allows also to detect operator errors of the type "wrong order". However, the pattern IMMEDIATELY NEXT would create wrong error messages for concurrent sequences and can only be used for non-concurrent activities.

## Pattern IMMEDIATELY NEXT
**Description:** This pattern describes pairs of activities, defining which activity has to be executed right after another activity (without any other activities occurring inbetween). The pattern is used for controlling XOR-joins, XOR-splits and non-concurrent sequences in a maintenance plan.

**Instances:** To get all instances of this pattern for a maintenance plan i we create a relation $P2_i := AM_i \times AM_i$ with the tuples $(a_{cur}, a_{nex})$ where,

- the corresponding transitions of the activities $a_{cur}$ and $ta_{nex}$ are connected through the same place, and
- the corresponding transitions cannot be executed concurrent to other transitions.

**Example:** In Fig. 5 instances of the pattern IMMEDIATELY NEXT are (a1, a2), (a1, a3), (a2, a4), (a3, a5), (a4, a6) and (a5, a7).

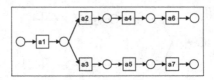

**Fig. 5.** Example pattern IMMEDIATELY NEXT

**Anti-pattern:** The anti-pattern allows to detect operator errors of the type "wrong order" for non-concurrent operations. An error message is created when a transaction event $te_{cur}$ occurs in the event stream and the next following transaction event $te_{cur+1}$ of the same service template does not correspond to the expected activity:

$$\pi_{app,op,prop} te_{cur} \in \pi_{a_{cur}.app,a_{cur}.op,a_{cur}.prop} P2_i \succ \pi_{app,op,prop}(\sigma_{st=te_{cur}.st} te_{cur+1})$$
$$\notin \pi_{app,op,prop}(\pi_{a_{nex}}(\sigma_{a_{cur}.app=te_{cur}.app \wedge a_{cur}.op=te_{cur}.op \wedge a_{cur}.prop=te_{cur}.prop} P2_i))$$

**Similar Pattern:** The pattern IMMEDIATELY NEXT is similar to the pattern NEXT. The difference is, that in the IMMEDIATELY NEXT pattern in contrast to the pattern NEXT no activities of the same service template are allowed between a pair of activities.

## Pattern PRECEDENCE
**Description:** This pattern describes pairs of activities where one activity has to occur before another one. Like the pattern NEXT it is allowed that other activities occur inbetween the activities of such a pair of activities. The pattern is used for controlling AND-joins, AND-splits and concurrent sequences in a maintenance plan.

**Instances:** To get all instances of the pattern a relation $P3_i := AM_i \times AM_i$ with the tuples $(a_{cur}, a_{pre})$ is created where,

- the corresponding transitions of the activities $a_{cur}$ and $a_{pre}$ are connected through the same place, and
- the corresponding transitions can be executed concurrent to other transitions.

**Example:** In Fig. 4 instances of the pattern PRECEDENCE are (a2, a1), (a3, a1), (a4, a2), (a5, a3), (a6, a4) and (a7, a5).

**Anti-pattern:** With the anti-pattern it is possible to detect operator errors of the type "wrong order". An error message is created when a transaction event $te$ occurs in the event stream which corresponds to an activity $a_{cur}$, but the corresponding transaction event for the activity $a_{pre}$ does not exist in the transaction event history:

$$\pi_{app,op,prop} te \in \pi_{a_{cur}.app,a_{cur}.op,a_{cur}.prop} P3_i$$
$$\wedge \pi_{app,op,prop}(\pi_{a_{pre}}(\sigma_{a_{cur}.app=te.app \wedge a_{cur}.op=te.op \wedge a_{cur}.prop=te.prop} P3_i))$$
$$\notin \pi_{app,op,prop} TEH$$

**Similar Pattern:** The pattern IMMEDIATELY PRECEDENCE is also used to detect forgotten activities, which have to be executed before another activity. For the pattern IMMEDIATELY PRECEDENCE no activities are allowed between $a_{cur}$ and $a_{pre}$, whereas for the pattern PRECEDENCE additional activities in between are allowed. Besides, the pattern is similar to the pattern NEXT. The difference is, that the pattern NEXT checks for future activities, whereas the pattern PRECEDENCE checks for activities happened in the past of a maintenance execution.

## Pattern IMMEDIATELY PRECEDENCE
**Description:** This pattern describes which activity has to be executed immediately before another one. It can be used for non-concurrent activities as well as for XOR-joins and XOR-splits.

**Instances:** To get all instances of this pattern for a maintenance plan i we create a relation $P4_i := AM_i \times AM_i$ with the tuples $(a_{cur}, a_{pre})$, where

- the corresponding transitions of the activities $a_{cur}$ and $a_{pre}$ are connected through the same place, and
- the corresponding transitions cannot be executed concurrent to other transitions.

**Example:** In Fig. 5 instances of the pattern IMMEDIATELY PRECEDENCE are (a2, a1), (a3, a1), (a4, a2), (a5, a3), (a6, a4) and (a7, a5).

**Anti-pattern:** With the anti-pattern it is possible to detect operator errors of the type "wrong order". An error message is created when a transaction event te occurs in the event stream which corresponds to an activity $a_{cur}$, but the latest transaction event of the same service template in the transaction event history does not correspond to $a_{pre}$:

$$\pi_{app,op,prop}te \in \pi_{a_{cur}.app,a_{cur}.op,a_{cur}.prop}P4_i$$
$$\wedge\, \pi_{app,op,prop}\big(\pi_{a_{pre}}(\sigma_{a_{cur}.app=te.app \wedge a_{cur}.op=te.op \wedge a_{cur}.prop=te.prop}P4_i)\big)$$
$$\notin \pi_{app,op,prop}\big(\sigma_{max(timestamp)}(\sigma_{timestamp<te.timestamp \wedge st=te.st})\big)TEH$$

**Similar Pattern:** The pattern PRECEDENCE describes also activities which have to occur before another activity. In contrast to the pattern IMMEDIATELY PRECEDENCE the pattern PRECEDENCE allows other activities of the same service template to occur between a pair of activities.

## Pattern STATE-CONDITION
**Description:** This pattern describes the state an application should have in order to be able to perform an operation on either the same or another application. Example: in order to shut down an application server, the database server must be in the state offline.

**Instances:** Instances of this pattern are all conditions $SC_i$ for a maintenance plan i.

**Example:** In Fig. 6, which is a snippet of Fig. 3, the instance of the pattern STATE-CONDITION is (MyApplication, start, NULL, TestDatabase, started).

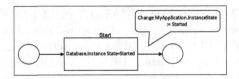

**Fig. 6.** Example pattern STATE-CONDITION

**Anti-pattern:** This anti-pattern does actually not detect an error like described in Sect. 2. Instead, it detects malicious prerequisites, which would lead to an operation error. This is done by comparing the latest state of an application with the planned state:

$$\pi_{app,op,prop}te \in \pi_{app,op,prop}SC_i$$

$$\wedge\ (\pi_{zapp,state}(\sigma_{app=te.app\wedge op=te.op\wedge prop=te.prop}SC_i)/\pi_{app,state}SEH) \neq \emptyset$$

**Similar Pattern:** There are no similar patterns for the STATE-CONDITION pattern.

**Pattern VALUE**
**Description:** This pattern describes the value of a configuration parameter which has to be changed during the maintenance.
**Instances:** To get all instances of this pattern a selection on the set of all activities of the maintenance plan is performed in order to get only those activities which include a change of a property: $P5 := \pi_{app,op,prop,value}\ (\sigma_{prop\neq NULL}SA_i)$.
**Example:** In Fig. 3 the only instance of this pattern is (MyApplication, configure, DB Connection, jdbc:mysql://localhost:1521/XE).
**Anti-pattern:** This anti-pattern allows to detect operator errors of the types "wrongly executed activity", "lexical error", "local inconsistency", "global inconsistency" and "typo" by checking the element value of a transaction event te:

$$\pi_{app,op,prop}ve \in \pi_{app,op,prop}P5 \wedge \pi_{app,op,prop,value}ve \notin P5_i$$

**Similar Pattern:** This pattern can be seen as a more detailed version of the OCCURRENCE pattern. However, the OCCURRENCE pattern just checks for executed operations and properties, but not for the actual values of modified properties.

**Pattern OCCURRENCE**
**Description:** This pattern includes all activities which have to be executed in a maintenance plan independent of the chosen path through the maintenance plan.
**Instances:** To get all instances of this pattern a set $P6_i$ with all activities of the maintenance plan i is created, where

- every activity has to be executed independent of the chosen path in the maintenance plan.

**Example:** In Fig. 3 the instances of this pattern are (MyService, MyApplication, stop, NULL, NULL), (MyService, MyApplication, configure, DB Connection, jdbc:mysql://localhost:1521/XE) and (MyService, MyApplication, start, NULL, NULL).

**Anti-pattern:** The anti-pattern allows to detect errors of the kind "forgotten activity". An error message is created, when an activity was executed too seldom:

$$\exists\ a \in \sigma_{time\ >\ a.end}P6_i \wedge \text{countTE}(\pi_{app,op,prop}a,$$
$$\pi_{app,op,prop}(\pi_{timestamp\geq a.start\ \wedge\ timestamp\leq a.ende}TEH))$$
$$< \text{countA}(\pi_{app,op,prop}a,\ \pi_{app,op,prop}(\pi_{start\geq a.start\ \wedge\ end\leq a.end}P6_i))$$

**Similar Pattern:** With the anti-patterns NEXT, IMMEDIATELY NEXT, PRECEDENCE and IMMEDIATELY PRECEDENCE it is also possible to detect forgotten activities in a limited way. However it is only possible to detect a forgotten activity right before or after another activity. As an example lets assume we have a sequence (a, b, c, d, e). If the activity a and e occur, it is possible to detect the forgotten activities b and d with the similar patterns, but not the activity c. Only with the anti-pattern OCCURRENCE it is possible to detect the forgotten activity c.

## Pattern ALTERNATIVE OCCURRENCE

**Description:** This pattern describes a pair of activities which cannot be executed together, like after XOR-Splits. However, one activity of such a pair of activities has to be performed during the maintenance.

**Instances:** To get all instances of this pattern a relation $P7_i := AM_i$ x $AM_i$ with the tuples $(a_{cur}, a_{alt})$, where

- the corresponding transitions of the activities $a_{cur}$ and $a_{alt}$ do not occur together in any path of the maintenance plan.

**Example:** In Fig. 5 instances of this pattern are (a2, a3), (a2, a5), (a2, a7), (a4, a3), (a4, a5), (a4, a7), (a6, a3), (a6, a5), (a6, a7), (a3, a2), (a3, a4), (a3, a6), (a5, a2), (a5, a4), (a5, a6), (a7, a2), (a7, a4) and (a7, a6).

**Anti-pattern[1]:** The anti-pattern allows to detect errors of the kind "forgotten activity". An error message is created, when activities of $P6_i$ were not executed. However, no error message is created for activities, if one alternative activity was already performed. The anti-pattern assumes, that the first executed alternative activity is the right one and therefore ignores all other activities, which may not be executed in conjunction with this first alternative activity.

**Similar Pattern:** The pattern OCCURRENCE does also detect forgotten activities, but it would create wrong error messages for exclusive activities, if already one of the exclusive activities was executed.

## Pattern ABSENCE

**Description:** This pattern describes which activities may not occur during a maintenance.

---

[1] Due to space limitations we forgo the formal definition of the following anti-patterns.

**Instances:** The instances of this pattern are all possible activities, which could really occur during a maintenance, without all activities, which are also modelled in the maintenance plan. Note that in a maintenance plan only a subset of possible operations on service templates is modelled and therefore planned. All other operations should not occur during the maintenance. The number of instances of this pattern can get very high, because the number of possible operations, especially configurations can be huge. However, for the anti-pattern of ABSENCE the generation of pattern instances of the type ABSENCE is not needed as it is explained in the following.

**Example:** On the assumption that in Fig. 3 other possible operations of *MyApplication* would be "deploy" and "undeploy", some of the instances of the pattern ABSENCE would be (MyService, MyApplication, deploy, NULL, NULL) and (MyService, MyApplication, undeploy, NULL, NULL).

**Anti-pattern:** The anti-pattern detects errors of the kind "unneeded activity". An error message is created either when

- a transaction event does not correspond to one of the activities in $P6_i$ or $P7_i$, or
- a transaction event corresponds to one the of the activities in $P6_i$ or $P7_i$, but it did not occur during the planned maintenance window, or
- a transaction event corresponds to one the of the activities in $P6_i$ or $P7_i$ and it occurred during the planned maintenance plan, but it occurred too often during the maintenance window.

**Similar Pattern:** With the anti-patterns NEXT, IMMEDIATELY NEXT, PRECEDENCE and IMMEDIATELY PRECEDENCE it is also possible to detect unneeded activities, when the following or precedence activity was not the planned one. However, these anti-patterns do not know, if the unneeded activity is an activity which was just executed in the wrong order or if it is an activity which should not occur at all.

### Pattern ALTERNATIVE ABSENCE

**Description:** This pattern describes activities which are not allowed to be executed depending on other activities. Such activities occur after XOR-Splits.

**Instances:** Instances of this pattern are the same like for the pattern ALTERNATIVE OCCURRENCE. Although the pattern instances are the same, the anti-pattern is different to the anti-pattern of ALTERNATIVE OCCURRENCE.

**Example:** For an example please see the examples of the pattern ALTERNATIVE OCCURRENCE.

**Anti-pattern:** An error message is created when one of the following conditions hold:

- A transaction event corresponds to an activity in $P7_i$. It is the first alternative activity and it occurred during the maintenance window, however it was performed too often.
- A transaction event corresponds to an activity in $P7_i$ and it occurred during the maintenance window, but an alternative activity was already performed before.

**Similar Pattern:** The pattern ABSENCE is similar, but the pattern does not check the absence of activities dependent of other activities.

### 4.3 Derivation of Pattern Instances

Pattern instances can be derived from the maintenance plan by simulating it. Therefore, the maintenance plan is marked with the Service Template of the to be maintained IT Service. The resulting simulation log is used to create log-based ordering relations and footprints like they are used in process mining and described in [18,19]. Based on these ordering relations two footprints are created. One footprint uses the basic ordering relations described in [18]. This footprint is used to derive the pattern instances IMMEDIATELY NEXT, PRECEDENCE, IMMEDIATELY PRECEDENCE and the activities $a_{cur}$ and $a_{nex}$ for the pattern instances of NEXT. In order to get $a_{far}$ for the pattern instances of NEXT the second footprint is used, which is based on the extended ordering relations described in [19].

Instances of the pattern ALTERNATIVE OCCURRENCE and ALTERNATIVE ABSENCE are also derived from the second footprint. The pattern OCCURRENCE is instantiated with all simulated activities, which occur in every path of the simulation log. For the anti-pattern ABSENCE all possible activities are needed, which can be derived directly from the simulation log.

Instances of the pattern STATE-CONDITION can be derived from the activities in the simulation log and the corresponding function defined in Definition 4.5. The pattern VALUE can be instantiated by filtering all activities in the simulation log, whose attribute *prop* is not NULL.

## 5   Implementation

The architecture of the proof of concept implementation consists of four main components and is shown in Fig. 7. The first component is a modelling component, which allows to model maintenance plans and derive pattern instances of a maintenance plan. The modelling component is implemented in the software tool Horus[2] and already allows to model generic XML nets. The extension of the tool in order to model TOSCA service templates and link them to an XML net is currently under construction.

The second main component are the log agents. Log agents are used to get every new log entry of an application, transform the log entry into the format of a transaction event and send it to the complex event processing engine. In the proof of concept log agents are implemented with Beats and Logstash[3]. Both products are developed for fast log data extraction. Besides, Logstash contains a powerful regular expression engine, which supports the transformation of proprietary log entries into the generic format of transaction events.

---

[2] www.horus.biz.
[3] https://elastic.co.

The third component is an IT infrastructure monitoring tool like Nagios[4], CloudWatch[5], or Metricbeat[6] which allows to check the state of an application in order to generate the state events. In the proof of concept we use Metricbeat.

The fourth component is the complex event processing engine, which checks incoming state and transaction events against the pattern instances of the maintenance plan. In the proof of concept the complex event processing system of WSO2[7] is used. All anti-patterns are implemented as event queries in the event pattern language Siddhi[8] and have to be implemented only once. In order to check future maintenance plans, only the corresponding pattern instances have to be transferred to the complex event processing system. As an example, for an anti-pattern written in Siddhi, see the following anti-pattern NEXT, implemented as Siddhi query:

```
from te [(app == NEXT.appcur and op == NEXT.opcur
and prop == NEXT.propcur) in NEXT] insert into #temp;
from #temp as t join NEXT as n on t.app == n.appcur and
t.op == n.opcur and t.prop == n.propcur
select t.timestamp, n.appcur, n.opcur, n.propcur, n.appnex,
n.opnex, n.propnex, n.appfar, n.opfar, n.propfar
insert into #temp1;
from e1=#temp1 -> e2= incoming_te [e1.appfar == e2.app
and e1.opfar == e2.op and e1.propfar == e2.prop]
select e1.timestamp, e1.appcur, e1.opcur, e1.propcur,
e1.appnex, e1.opnex, e1.propnex, e2.timestamp as timestampfar
insert into #temp2;
from #temp2 [not((appnex == TEH.app and opnex == TEH.op
and propnex == TEH.prop in and timestamp < TEH.timestamp
and timestampentf >  TEH.timestamp) in TEH)]
select str:concat("The activity ",appnex, ", ", opnex, ", "
, propnex, " was not performed after the activity ", appcur,
" ,", opcur, ", ", propcur, ".") as message
insert into error_message;
```

Apart of the modelling component all components and Siddhi queries are implemented in a prototype, which is used to evaluate the approach. A first evaluation experiment was conducted, which is described in the following.

---

[4] https://nagios.org.
[5] https://aws.amazon.com/en/cloudwatch/.
[6] https://www.elastic.co/guide/en/beats/metricbeat/6.2/index.html.
[7] https://wso2.com/products/complex-event-processor/.
[8] https://github.com/wso2/siddhi.

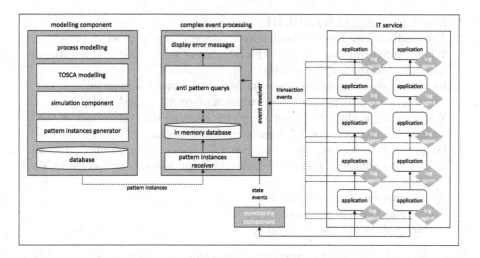

**Fig. 7.** Implementation architecture

## 6 Experimental Results

To evaluate the approach an exemplary IT service maintenance was performed. Therefore, we built an IT service environment using Amazon EC2[9] machines. The environment contains five EC2 machines. On two machines we installed an Apache webserver[10] hosting the open source application SugarCRM[11]. On two other machines Maria DB was installed. Both SugarCRM instances connect to the same Maria DB instance. In the experiment the configuration of both SugarCRM instances should be changed, so SugarCRM connects only to the second Maria DB instance anymore. The fifth EC2 machine is used to host the complex event processing engine and Logstash in order to transform log data and check the data for anti-patterns. Figure 8 gives an overview of the maintenance plan, which was used for evaluation and to derive the pattern instances for the experiment. The maintenance plan is described shortly in the following.

According to the maintenance plan first of all the database server named database2 needs to be started. Afterwards the loadbalancer is shut off in order to avoid connections to the webservers. When the loadbalancer is offline, the two webservers have to be stopped and reconfigured in order to connect to database2. After reconfiguring and stopping the webservers, they can be started again. If both webservers where stopped, the former database server can be shut down. Finally, the loadbalancer ha to be started again in order to redirect requests to the webservers. During the execution of this maintenance plan typical operator errors, like they are described in Sect. 2, were injected. Namely, those operator errors are:

---

[9] https://aws.amazon.com/ec2/.
[10] https://httpd.apache.org.
[11] https://www.sugarcrm.com.

1. Forgot to configure SugarCRM1
2. The loadbalancer was started before webserver2 was started
3. When configuring SugarCRM1 the IP address is changed and additionally without need the user is modified
4. Instead of stopping databaseserver1, databaseserver2 is stopped
5. The wrong IP address is entered, when configuring SugarCRM1
6. When changing the IP address a typo happens, so that the format of the IP address is xxx.xxx.xxxxxx instead of xxx.xxx.xxx.xxx
7. Webserver2 is started. However, the EC2 machine is stopped manually in order to simulate a software bug which hinders the webserver to start properly
8. A combination of error 1 and 2
9. A combination of error 3 and 4
10. A combination of error 3, 4 and 6.

In the first ten runs of the experiment one error was injected per run. Afterwards we repeated the experiment. However, in the second ten runs, errors were corrected immediately after their detection. By correcting them, the actual maintenance execution differs from the maintenance plan, because we did not model any procedural exception handlings.

In order to quantify the results, we used the metrics *Recall*, *Precision* and *F-Score* known from machine learning evaluations [24].

In the first ten runs all errors were identified and no false positives were reported, resulting in a Precision and Recall of 100%. The negative site of this is, that error messages were created multiple times for the same root cause. For example when the configuration of SugarCRM1 was forgotten (error 1) the anti-pattern NEXT as well as the anti-pattern PRECEDENCE created an error message, when the activity after the forgotten activity was executed. Additionally, the anti-pattern OCCURRENCE created an error message, because the occurrence of the activity could not be found in the transaction event history. In some runs this led to a ratio of up to four created error messages for one root cause.

In the second round of the experiment, when the errors were corrected after their identification, the precision decreased to 41%. The reason was the increasing number of false positives. For example when the forgotten configuration of SugarCRM1 was identified, the error was corrected by stopping webserver1 again, configuring SugarCRM1 and starting webserver1 again. When webserver1 was stopped respectively started the anti-pattern ABSENCE created two error messages that webserver1 was stopped respectively started too many times. Nonetheless, all injected errors were identified resulting in a Recall of 100%. Besides, the ratio of reported error messages to root causes improved, because of the immediate error handling. The F-Score over all runs is 73%. The F-Score of the first ten runs is 100%, whereas the F-Score only for the runs with immediate error handling is 58%.

In the next months we plan to conduct additional experiments to test the recall and precision of the method. Besides, we plan to perform performance

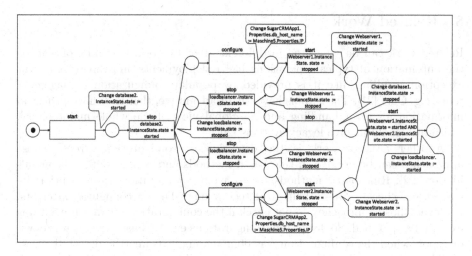

**Fig. 8.** Maintenance plan used for the experiment

tests as this is one of our main objectives, that the method reports errors within seconds. So, operators would have a realistic chance to correct their errors before they manifest in an IT service.

## 7 Discussion

The high false positive rate when error handling was performed can be interpreted as overfitting. On the contrary, the false positive rate would decrease, if error handling would be modelled explicitly in the maintenance plan. However, we think it is quite unlikely that an operator would model every possible exception, because this could lead to a very complex unreadable maintenance plan.

In general checking conformance of an event stream to a process model in an online setting leads to new challenges, that do not exist for traditional offline conformance checking methods. In offline conformance checking methods, analyzing event traces instead of event streams, the actual process execution can not be influenced anymore (apart of long running process, which are not finished when analyzing the log trace). In an online setting like in this work an operator would adjust the process execution spontaneous, because of the identification of errors or deviations from the process modell. This could lead to process executions differing a lot from the actual modelled process.

An error detection method or conformance checking method should be able to recognize, which deviations from the process modell are allowed ones, e.g. because of correcting an error and which one are real errors in order to reduce false positives. Therefore, we plan to extend our approach by adding a machine learning component, which analyzes which error messages are real errors and which were created because of correcting an already performed error.

# 8    Related Work

Related work can be separated in different areas of work. One area of work is the automation of typical operations like redeployments and integrated error exception handling, like it is provided by popular configuration management tools, e.g. Chef [21]. Those tools have the disadvantage, that they have just local information for error handling and no global view of the whole maintenance, which also could involve legacy systems [20].

Another area of work is the detection of configuration errors. Those approaches can be divided in rule based methods and online configuration validation [22]. Rule based methods try to avoid configuration errors a priori by correctness checks. These, help to detect wrong planned configuration errors. However, those approaches do not check if the configuration operation itself was executed as planned. So, forgotten configurations e.g. because a server was down or typos, when the configuration was done manually, cannot be detected.

The most related work to ours is the work of Xu et al. [20] and Farshchi et al. [23]. Both works describe an approach to monitor sporadic operations in cloud environments. Xu et al. developed a method called "POD-Diagnosis". They use a process model to detect operator errors through token replay by checking the conformance of observed logs with the prebuild model and an additional fault tree analysis in order to find the root cause of the error. In contrast to our work only the control flow of the process is modelled and can therefore be checked. Apart of that, in our approach no additional fault tree has to be build. Farshchi et al. build a regression-based model to find correlation and causalities between events described in logs and overserved metrics of resources. In their approach, assertions are derived from the regression-based model. However, they are also limited to control flow. Additionally, enough learning data is needed, which practically limits their approach to automated cloud environments. Our approach does not have to learn data and therefore can also be used to monitor manually executed steps or changes in legacy systems as long as those actions can be seen in the logs of the systems.

The field of business process compliance monitoring can also be seen as other related work, which uses patterns to check that process executions adhere to predefined compliance requirements [26]. Our work differs from this field in the way, that we do not use patterns to make sure, that specific compliance requirements like the "segregation of duty" pattern are covered during a process execution. The patterns used in this work are only used in order to be able to instantiate anti-patterns during process executions in order to identify situations, which must not occur. The notion of anti-pattern can be seen as the counterpart to compliance pattern [26]. The most related work to ours in this field is [27], who also use the notion of anti-patterns instead of compliance patterns. However, in our work no definition of patterns has to be performed by an user. Instances of the patterns are derived automatically from the created maintenance plan during design phase, which integrates the control flow and configuration modelling. The instantiation of patterns can even be influenced after modelling by the simulation of the maintenance plan. If paths of the maintenance plan will not be simulated, those will not be represented in the simulation log and therefore cannot be used for pattern instantiation. During the execution phase anti-patterns check only deviations from the derived pattern instances.

## 9 Conclusion

In this paper, we describe an approach to detect operator errors online during the execution of maintenance operations. Therefore, we define different anti-patterns, which are implemented as complex event processing queries and check in real time log entries and state metrics of observed resources against pattern instances of a predefined process model. The process model itself is realized as a TOSCA based XML net, combining the modelling of the control-flow and the resources. A first evaluation with a prototype implementation was performed, resulting in a very good error detection rate. On the contradictory site, the approach can result in a high false positive rate, when the process execution is adjusted spontaneously in order to correct the reported errors. Therefore, we plan to extend our approach in order to deal with this spontaneous flexibility during an IT service maintenance.

## References

1. Gunawi, H.S., et al.: What bugs live in the cloud? A study of 3000+ issues in cloud systems. In: Proceedings of the ACM Symposium on Cloud Computing, pp. 1–14 (2014)
2. Hagen, S., Seibold, M., Kemper, A.: Efficient verification of IT change operations or: how we could have prevented Amazon's cloud outage. Presented at the Network Operations and Management Symposium (NOMS), 2012 IEEE, pp. 368–376 (2012)
3. Dumitra, T., Narasimhan, P.: Why do upgrades fail and what can we do about it? Toward dependable, online upgrades in enterprise system. In: Proceedings of the 10th ACM/IFIP/USENIX International Conference on Middleware, p. 18 (2009)
4. Pertet, S., Narasimhan, P.: Causes of failure in web applications. Parallel Data Laboratory, p. 48 (2005)
5. Oppenheimer, D., Ganapathi, A., Patterson, D.A.: Why do internet services fail, and what can be done about it? In: Proceedings of the 4th Conference on USENIX Symposium on Internet Technologies and Systems, vol. 4, Berkeley, p. 1 (2003)
6. Scott, D.: Making smart investments to reduce unplanned downtime. Tactical Guidelines Research Note TG-07-4033, Gartner Group, Stamford, CT (1999)
7. Elliot, S.: DevOps and the cost of downtime: fortune 1000 best practice metrics quantified. International Data Corporation, IDC (2014)
8. Vetter, A.: Detecting operator errors in cloud maintenance operations. In: 2016 IEEE International Conference on Cloud Computing Technology and Science (CloudCom), pp. 639–644 (2016)
9. Nagaraja, K., Oliveira, F., Bianchini, R., Martin, R.P., Nguyen, T.D.: Understanding and dealing with operator mistakes in internet services. In: OSDI 2004: 6th Symposium on Operating Systems Design and Implementation (2004)

10. Yin, Z., Ma, X., Zheng, J., Zhou, Y., Bairavasundaram, L.N., Pasupathy, S.: An empirical study on configuration errors in commercial and open source systems. In: Proceedings of the Twenty-Third ACM Symposium on Operating Systems Principles, pp. 159–172 (2011)
11. Peterson, J.L.: Petri Net Theory and the Modeling of Systems. Prentice Hall, Upper Saddle River (1981)
12. Gamma, E., Helm, R., Johnson, R., Vlissides, J.: Design Patterns: Elements of Reusable Object-Oriented Software. Pearson Education, London (1994)
13. van der Aalst, W.M., Ter Hofstede, A.H., Kiepuszewski, B., Barros, A.P.: Workflow patterns. Distrib. Parallel Databases 14(1), 5–51 (2003)
14. Russell, N., Ter Hofstede, A.H., Edmond, D., van der Aalst, W.M.: Workflow Data Patterns. QUT Technical report, FIT-TR-2004-01. Queensland University of Technology, Brisbane (2004)
15. Russell, N., van der Aalst, W.M.P., ter Hofstede, A.H.M., Edmond, D.: Workflow resource patterns: identification, representation and tool support. In: Pastor, O., Falcão e Cunha, J. (eds.) CAiSE 2005. LNCS, vol. 3520, pp. 216–232. Springer, Heidelberg (2005). https://doi.org/10.1007/11431855_16
16. Riehle, D., Züllighoven, H.: Understanding and using patterns in software development. TAPOS 2(1), 3–13 (1996)
17. Dwyer, M.B., Avrunin, G.S., Corbett, J.C.: Property specification patterns for finite-state verification. In: Proceedings of the Second Workshop on Formal Methods in Software Practice, pp. 7–15 (1998)
18. Van Der Aalst, W.: Process Mining: Discovery, Conformance and Enhancement of Business Processes. Springer, Heidelberg (2011). https://doi.org/10.1007/978-3-642-19345-3
19. Weidlich, M., Mendling, J., Weske, M.: Computation of behavioural profiles of process models. Business Process Technology, Hasso Plattner Institute for IT-Systems Engineering, Potsdam (2009)
20. Xu, X., Zhu, L., Weber, I., Bass, L., et al.: POD-diagnosis: error diagnosis of sporadic operations on cloud applications. In: 2014 44th Annual IEEE/IFIP International Conference on Dependable Systems and Networks, pp. 252–263 (2014)
21. Chef: About Handlers, 08 November 2017. https://docs.chef.io/handlers.html
22. Xu, T., Zhou, Y.: Systems approaches to tackling configuration errors: a survey (2014)
23. Farshchi, M., Schneider, J.-G., Weber, I., Grundy, J.: Metric selection and anomaly detection for cloud operations using log and metric correlation analysis. J. Syst. Softw. 137, 531–549 (2017)
24. Powers, D.: Evaluation: from precision, recall and F-measure to ROC, informedness, markedness and correlation. J. Mach. Learn. Technol. 2(1), 37–63 (2011)
25. Kopp, O., Binz, T., Breitenbücher, U., Leymann, F.: BPMN4TOSCA: a domain-specific language to model management plans for composite applications. In: Mendling, J., Weidlich, M. (eds.) BPMN 2012. LNBIP, vol. 125, pp. 38–52. Springer, Heidelberg (2012). https://doi.org/10.1007/978-3-642-33155-8_4
26. Becker, M., Klingner, S.: A criteria catalogue for evaluating business process pattern approaches. In: Bider, I., et al. (eds.) BPMDS/EMMSAD-2014. LNBIP, vol. 175, pp. 257–271. Springer, Heidelberg (2014). https://doi.org/10.1007/978-3-662-43745-2_18
27. Awad, A., Barnawi, A., Elgammal, A., Elshawi, R., Almalaise, A., Sakr, S.: Runtime detection of business process compliance violations: an approach based on anti patterns. In: 12th Enterprise Engineering Track at ACM, SAC 2015 (2015)

# Executing Lifecycle Processes in Object-Aware Process Management

Sebastian Steinau$^{(\boxtimes)}$, Kevin Andrews, and Manfred Reichert

Institute of Databases and Information Systems, Ulm University, Ulm, Germany
{sebastian.steinau,kevin.andrews,manfred.reichert}@uni-ulm.de

**Abstract.** Data-centric approaches to business process management, in general, no longer require specific activities to be executed in a certain order, but instead data values must be present in business objects for a successful process completion. While this holds the promise of more flexible processes, the addition of the data perspective results in increased complexity. Therefore, data-centric approaches must be able to cope with the increased complexity, while still fulfilling the promise of high process flexibility. Object-aware process management specifies business processes in terms of objects as well as their lifecycle processes. Lifecycle processes determine how an object acquires all necessary data values. As data values are not always available in the order the lifecycle process of an object requires, the lifecycle process must be able to flexibly handle these deviations. Object-aware process management provides operational semantics with built-in flexible data acquisition, instead of tasking the process modeler with pre-specifying all execution variants. At the technical level, the flexible data acquisition is accomplished with process rules, which efficiently realize the operational semantics.

**Keywords:** Lifecycle execution · Data-centric processes · Flexible data acquisition · Process rules

## 1 Introduction

Data-centric modeling paradigms part with the activity-centric paradigm, and instead base process modeling and enactment on the acquisition and manipulation of business data. In general, a data-centric process no longer requires certain activities to be executed in a specific order for successful completion. Instead certain data values must be present, regardless of the order in which they are acquired. Activities and decisions consequently rely on data conditions for enactment, e.g., an activity becomes executable once required data values are present. While this holds the promise of vastly more flexible processes in theory, it is no sure-fire success. The increased complexity from considering the data perspective in addition to the control-flow perspective requires a thoughtful design of any approach for modeling and enacting data-centric processes. This design should enable the flexibility of data-centric processes, while still being able to manage the increased complexity.

© IFIP International Federation for Information Processing 2019
Published by Springer Nature Switzerland AG 2019
P. Ceravolo et al. (Eds.): SIMPDA 2017, LNBIP 340, pp. 25–44, 2019.
https://doi.org/10.1007/978-3-030-11638-5_2

Object-aware process management [16] is a data-centric approach to business process support that aims to address this challenge. In the object-aware approach, business data is held in *attributes*. Attributes are grouped into *objects*, which represent logical entities in real-world business processes, e.g., a loan application or a job offer. Each object has an associated *lifecycle process* that describes which attribute values need to be present for successfully processing the object. Lifecycle processes adopt a modeling concept that resembles an imperative style, i.e., the model specifies the default order in which attribute values are required. Studies have indicated that imperative models show advantages concerning understandability compared to declarative models, which are known for flexibility [11,19,20]. While the imperative style allows for an easy modeling of lifecycle processes, it seemingly subverts the flexibility promises of the data-centric paradigms, as imperative models tend to be rather rigid [25]. However, in object-aware process management, the operational semantics of lifecycle processes allow data to be entered at any point in time, while still ensuring correct process execution. The imperative model provides only the basic structure. This has the advantage that modelers need not concern themselves with modeling flexible processes, instead the flexibility is built into the operational semantics of lifecycle processes.

The functional specifications of the operational semantics of lifecycle processes have partially been presented in previous work [15]. This paper expands upon this work and contributes extended functionality and the technical implementation of the operational semantics, provided in the PHILharmonicFlows prototype. In particular, the logic involving execution events has been completely redesigned to include completion and invalidation events. These event types became necessary as otherwise the consistency of the lifecycle process was not guaranteed. Further, decision making in lifecycle processes has been improved by redesigning the data-driven operational semantics of decisions.

The technical implementation is based on the *process rule framework*, a lightweight, custom process rule engine. The framework is based on event-condition-action (ECA) rules, which enable reacting to every contingency the functional specification of the operational semantics permit, i.e., correct lifecycle process execution is ensured. The process rule framework will further provide the foundation for implementing the operational semantics of *semantic relationships* and *coordination processes*, the object-aware concept for coordinating objects and their lifecycle processes [23]. Such a coordination is necessary, as objects interact and thereby form large process structures, constituting an overall business process [22]. As such, coordination processes enable collaborations of concurrently running lifecycle processes, having the advantage of separating lifecycle process logic and coordination logic. With the transition of PHILharmonicFlows to a hyperscale architecture [2], the process rule framework is fully compatible with the use of microservices, enabling a highly concurrent execution of multiple lifecycle processes with large numbers of user interactions.

The remainder of the paper is organized as follows: Sect. 2 provides the fundamentals of object-aware process management. In Sect. 3, the extended operational semantics are presented. The process rule framework at the core of the operational semantics implementation is described in Sect. 4. Finally, Sect. 5 discusses related work, whereas Sect. 6 concludes the paper with a summary and an outlook.

## 2   Fundamentals

Object-aware process management organizes business data in form of objects, which comprise attributes and a lifecycle process describing object behavior. PHILharmonicFlows is the implementation of the object-aware concept to process management. Object-aware process management distinguishes design-time entities, denoted as *types* (formally$^T$), and run-time entities, denoted as *instances* (formally$^I$). Collectively, they are referred to as entities. At run-time, types may be instantiated to create one or more corresponding instances. For the purposes of this paper, object instance (cf. Definition 1) and lifecycle process instance (cf. Definition 2) definitions are required. The corresponding type definitions can be found in [16]. The "dot" notation is used to describe paths, e.g., for accessing the name of an object instance. $\perp$ describes the undefined value.

**Definition 1.** *(Object Instance)*
*An **object instance** $\omega^I$ has the form $(\omega^T, n, \Phi^I, \theta^I)$ where*

- $\omega^T$ *refers to the object type from which this object instance has been generated.*
- $n$ *is the name of the object instance.*
- $\Phi^I$ *is a set of attribute instances $\phi^I$, where $\phi^I = (n, \kappa, v_\kappa)$, with $n$ as the attribute instance name, $\kappa$ as the data type (e.g., String, Boolean, Integer), and $v_\kappa$ as the typed value of the attribute instance.*
- $\theta^I$ *is the lifecycle process (cf. Definition 2) describing object behavior.*

An object's lifecycle process (cf. Definition 2) is responsible for acquiring data values for the attributes of the object.

**Definition 2.** *(Lifecycle Process Instance)*
*A **lifecycle process instance** $\theta^I$ has the form $(\omega^I, \Sigma^I, \Gamma^I, T^I, \Psi^I, E_\theta, \mu_\theta)$ where*

- $\omega^I$ *refers to the object instance to which this lifecycle process belongs.*
- $\Sigma^I$ *is a set of **state instances** $\sigma^I$, with $\sigma^I = (n, \Gamma_\sigma^I, T_\sigma^I, \Psi_\sigma^I, \mu_\sigma)$ where*
  - $n$ *is the state name.*
  - $\Gamma_\sigma^I \subset \Gamma^I$ *is subset of steps $\gamma^I$.*
  - $T_\sigma^I \subset T^I$ *is a subset of transitions $\tau^I$.*
  - $\Psi_\sigma^I \subset \Psi^I$ *is a subset of backwards transitions $\psi^I$.*
  - $\mu_\sigma$ *is the state marking.*
- $\Gamma^I$ *is a set of **step instances** $\gamma^I$, with $\gamma^I = (\phi^I, \sigma^I, T_{in}^I, T_{out}^I, P^I, \lambda, \mu_\gamma, d_\gamma)$ where*
  - $\phi^I \in \omega^I.\Phi^I$ *is an optional reference to an attribute instance $\phi^I$ from $\Phi^I$ of object instance $\omega^I$. Default is $\perp$.*
    *If $\phi^I = \perp$, the step is denoted as an **empty step instance**.*

- $\sigma^I \in \Sigma^I$ is the state instance to which this step instance $\gamma^I$ belongs.
- $T_{in}^I \subset T_\sigma^I$ is the set of incoming transition instances $\tau_{in}^I$.
- $T_{out}^I \subset T_\sigma^I$ is the set of outgoing transition instances $\tau_{out}^I$.
- $P^I$ is a set of **predicate step instances** $\rho^I$, $P^I$ may be empty, with $\rho^I = (\gamma^I, \lambda)$ where
  * $\gamma^I$ is a step instance.
  * $\lambda$ is an expression representing a decision option.
  If $P^I \neq \emptyset$, the step instance $\gamma^I$ is called a **decision step instance**.
- $\lambda$ is an optional expression representing a computation.
  If $\lambda \neq \bot$, the step instance $\gamma^I$ is called a **computation step instance**.
- $\mu_\gamma$ is the step marking, indicating the execution status of $\gamma^I$.
- $d_\gamma$ is the step data marking, indicating the status of $\phi^I$.

- $T^I$ is a set of **transition instances** $\tau^I$, with $\tau^I = (\gamma_{source}^I, \gamma_{target}^I, ext, p, \mu_\tau)$ where
  - $\gamma_{source}^I \in \Gamma$ is the source step instance.
  - $\gamma_{target}^I \in \Gamma$ is the target step instance.
  - $ext := \gamma_{source}^I.\sigma^I = \gamma_{target}^I.\sigma^I$ is a computed property, denoting the transition as external, i.e., it connects steps in different states.
  - $p$ is an integer signifying the priority of the transition.
  - $\mu_\tau$ is the transition marking.

- $\Psi^I$ is a set of **backwards transition instances** $\psi^I$, $\Psi^I$ may be empty, with $\psi^I = (\sigma_{source}^I, \sigma_{target}^I, \mu_\psi)$ where
  - $\sigma_{source}^I \in \Sigma^I$ is the source state instance.
  - $\sigma_{target}^I \in \Sigma^I$ is the target state instance, $\sigma_{target}^I \in Predecessors(\sigma_{source}^I)$.
  - $\mu_\psi$ is the backwards transition marking.

- $E_\theta$ is the event storage for $\theta^I$, storing execution events $\epsilon^E$.
- $\mu_\theta$ is the lifecycle process marking.

All sets are finite and must not be empty unless specified otherwise. The function $Predecessors: \sigma^I \to \Sigma^I$ determines a set of states from which $\sigma^I$ is reachable. The function $Successors$ is defined analogously.

Note that for the sake of brevity the value of a step $\gamma^I$ refers to the value of the corresponding attribute $\gamma^I.\phi^I$. Furthermore, correctness criteria have been omitted from Definitions 1 and 2. For the sake of clarity, a lifecycle process is described by a directed acyclic graph with one start state and at least one end state. Figure 1 shows object instance *Bank Transfer* with its attributes and lifecycle process. The object instance represents a simplified transfer of money from one account to another. For this purpose, the states and the steps of a lifecycle process can be used to *automatically generate forms*. This is unique for process management systems, as in other systems, forms must still be designed manually, leading to a huge difference regarding productivity [25]. Additionally, when executing a process, the auto-generated forms are filled in by authorized users. The PHILharmonicFlows authorization system and its connection to form auto-generation has been discussed in [1]. In essence, forms may be personalized automatically based on the user's permissions, no different form designs showing different form fields are necessary.

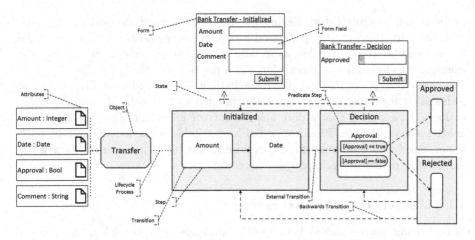

**Fig. 1.** Example object and lifecycle process of a Transfer

As depicted in Fig. 1, the state $\sigma^I_{Initialized}$ contains steps $\gamma^I_{Amount}$ and $\gamma^I_{Date}$, signifying that values for attributes $\phi^I_{Amount}$ and $\phi^I_{Date}$ are required during process execution. For the sake of brevity, the properties of an entity (e.g., the name of a step $\gamma$) may be written as a subscript, e.g., $\gamma_{Amount}$ for the first step in Fig. 1. The form corresponding to $\sigma^I_{Initialized}$ contains input fields for steps $\gamma^I_{Amount}$ and $\gamma^I_{Date}$. This means a state represents a form, whereas the steps represent form fields. The $\phi^I_{Comment}$ field is an optional field visible to a user due to the authorization system of PHILharmonicFlows. In state $\sigma^I_{Decision}$, a decision step $\gamma^I_{Approval}$ represents the approval of the bank for the money transfer. The automatically generated form displays $\gamma^I_{Approval}$ as a drop-down field. End states $\sigma^I_{Approved}$ and $\sigma^I_{Rejected}$ display an empty form, as the contained steps are empty (cf. Definition 2). Transitions determine at run-time which attribute value is required next, an external transition also determines the next state. Backwards transitions allow returning to a previous state, e.g., to correct a data value.

## 3   Lifecycle Process Operational Semantics

Data acquisition in PHILharmonicFlows is achieved through forms, which can be auto-generated from lifecycle process models $\theta^I$. A form itself is mapped to a state $\sigma^I$ of the lifecycle process $\theta^I$; form fields are mapped to steps $\gamma^I$. In consequence, the operational semantics of lifecycle processes emulate the behavior of electronic and paper-based forms, following a "best of both worlds" approach. Paper-based forms provide a great overview over the form fields, i.e., every form field may be viewed at any point in time. Further, they provide a reasonable default structure, but allow filling form fields at any point in time and in any order, e.g., starting to fill in form fields in the middle of the form is possible. In turn, electronic forms usually provide less overview, i.e., viewing subsequent forms is not possible before having filled out all mandatory fields in the current form. In contrast to paper-based forms, however, electronic forms are able to only display relevant fields, especially in context of decision branching. For example, an electronic anamnesis form at a physician's office may skip the questions related to pregnancy entirely if the patient is male. Additionally, electronic forms allow for data values to be easily changed as well as for data input verification, e.g., ensuring that a date has the correct format or all mandatory form fields possess a value. PHILharmonicFlows combines the advantages of both paper-based and electronic forms, providing flexibility in entering data while ensuring a correct lifecycle process execution.

### 3.1   Lifecycle Process Execution

For realizing the combined benefits, the progress of a lifecycle process $\theta^I$ is determined by its active state $\sigma_A^I$, i.e., marking $\sigma^I.\mu_\sigma = Activated$. Only one state $\sigma^I$ of $\theta^I$ may be active at any point in time. Per default, the form of the active state is displayed to a user when executing lifecycle process $\theta^I$. However, the user may choose to display forms of other states. When processing $\theta^I$, the active state changes, depending on data availability and decision results. For example, in regard to Fig. 1, starting the execution of the lifecycle process activates $\sigma_{Initialized}^I$. If values for steps $\gamma_{Amount}^I$ and $\gamma_{Date}^I$ are available (cf. Sect. 3.2), $\sigma_{Initialized}^I$ may be marked as $\mu_\sigma = Confirmed$, and the next state $\sigma_{Decision}^I$ becomes active, i.e., $\sigma_{Decision}^I.\mu_\sigma = Activated$. Depending on the value of $\gamma_{Approval}^I$, either $\sigma_{Approved}^I$ or $\sigma_{Rejected}^I$ becomes active. As both states are end states, the execution of $\theta^I$ terminates. The active state possesses a crucial role in the execution of $\theta^I$, as consequences from data acquisition or decisions are only evaluated for the active state. For example, providing value $true$ to $\gamma_{Approval}^I$ does not trigger the decision, if $\sigma_{Initialized}^I$ is the currently active state. This is to avoid inconsistent processing states, e.g., because a previous decision may make filling out a state $\sigma^I$ obsolete due to dead-path elimination [16].

For several reasons, including automatic form generation and process lifecycle coordination, only exactly one state may be active at a given point in time. If two or more state had become simultaneously active, it would be unclear which form should be presented to the user, or what the progress of the lifecycle is. State execution (cf. Sect. 3.2) must therefore enforce that only exactly one state may activate at the conclusion of a previous one. In consequence, the enabling of external transitions must be mutually exclusive. Regarding decisions steps and its predicate steps, additional measures are required to prevent the simultaneous enabling of different transitions.

For states $Successors(\sigma_A^I)$, data values may be entered, but processing only occurs once a state becomes active. All successor states possess marking $\mu_\sigma = Waiting$. If a user enters values for steps $\gamma^I$, these values will be stored and taken into account if the corresponding state $\gamma^I.\sigma^I$ becomes active. To indicate the status of the corresponding attribute value, steps possess a *data marking* $d_\gamma$. When setting the data value for a step $\gamma_{hasValue}^I$, where the state instance $\sigma^I$ has

$\mu_\sigma = Waiting$, the data marking of $\gamma^I_{hasValue}$ is set to $d_\gamma = Preallocated$. Should $\sigma^I$ become active during process execution, $d_\gamma = Preallocated$ will indicate that a value is present and thus is not be required anymore (cf. Sect. 3.2).

States that have already been processed, i.e., $Predecessors(\sigma^I_A)$, will either have marking $\mu_\sigma = Confirmed$ or $\mu_\sigma = Skipped$. States with marking $\mu_\sigma = Confirmed$ have previously been active, whereas skipped states have undergone a dead-path elimination. For reasons of data integrity, the values of steps in skipped or confirmed states must not be altered at any point in time. If allowed, inconsistencies and unpredictable execution behavior may occur. For example, changing values of decisions steps in an uncontrolled way might activate currently eliminated states, whereas currently active states become eliminated. However, it must be possible to correct mistakes for previously entered and accidentally confirmed data. Therefore, *backwards transitions* (cf. Definition 2) allow for the reactivation of confirmed states in a controlled way, where the data may be altered in a consistent and safe way; consequently, subsequent changes in decisions can be handled properly. The reactivation of states and correction of mistakes contributes much to the flexibility of object-aware lifecycle process execution.

### 3.2   State Execution

While PHILharmonicFlows is capable of auto-generating forms from states and steps, so far, these forms are static. However, there are dynamic aspects to a form, e.g., the indication which value is required next or which external transition or backwards transition may be committed. For this purpose, a lifecycle process $\theta^I$ provides *execution events* $\epsilon^E$ and an *event storage* $E_\theta$. Execution events are dynamically created when processing a lifecycle process $\theta^I$. When auto-generating a form, the static form is enriched with dynamic information from $E_\theta$ and displayed to the user. Execution events have different subtypes, namely *request events, completion events,* and *invalidation events.* When request events are created, they are stored in $E_\theta$ and are then used to enrich the form. Completion and invalidation events remove request events from $E_\theta$, when a request event are either fulfilled or no longer valid, respectively. The usage of the event storage $E_\theta$, in conjunction with the generated static forms, allows multiple users access to the same form, due to the centralized storage of the dynamic form data. The use of $E_\theta$ further allows preserving dynamic data over multiple sessions, i.e., a user may partially fill out a form, close it and do something else, and later return and continue where the user previously stopped. It is even possible that another user finishes filling out the form, introducing additional flexibility. In general, storing execution events $\epsilon^E$ ensures consistency regardless of any user interaction with the forms.

The creation and removal of execution events is primarily determined by the respective marking $\mu$ of states, steps, transitions, and backwards transitions. For steps with an attribute (i.e., $\gamma^I.\phi^I \neq \bot$), data marking $d_\gamma$ is also taken into account. For example, if step $\gamma^I_{Amount}$ in Fig. 1 has marking $\mu_\gamma = Enabled$, but $\gamma^I_{Amount}.d_\gamma = Unassigned$ holds, an *"attribute value request"* event is created and stored in $E_\theta$ after some intermediate processing steps. If a user accesses the form for $\sigma^I_{Initialized}$, the form field for $\gamma^I_{Amount}$ is tagged with an asterisk, indicating that a value is mandatory (cf. Fig. 2). As soon as the user provides a value for the $\gamma^I_{Amount}$ form field, the data marking for $\gamma^I_{Amount}$ is updated to $d_\gamma = Assigned$. This indicates that a value has been successfully provided for $\gamma^I_{Amount}$. In consequence, the attribute value request event in $E_\theta$ is no longer necessary. Therefore, setting $d_\gamma = Assigned$ triggers a completion event removing the "attribute value request" event from $E_\theta$. After the completion event has occurred, more markings change in a cascading fashion, leading to the step $\gamma^I_{Amount}$ being marked as $Unconfirmed$. This enables the outgoing transitions $\gamma^I_{Amount}$, which, in turn, leads to the next step $\gamma^I_{Data}$ receiving $\mu_\gamma = Enabled$. The data marking $\gamma^I_{Date}.d_\gamma = Unassigned$ triggers the same chain of events and marking changes analogously to the marking change of $\gamma^I_{Amount}$.

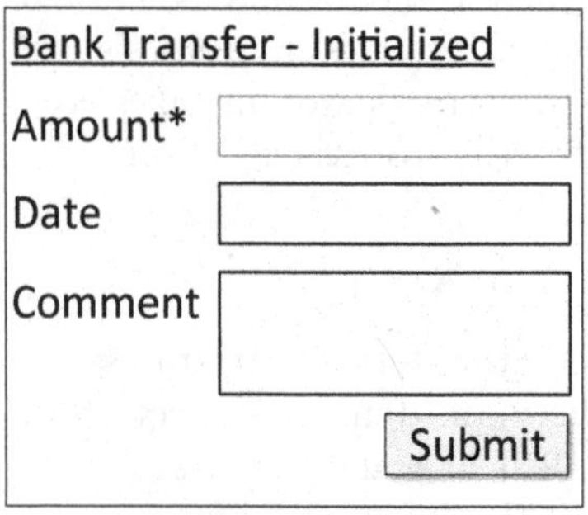

Fig. 2. Form enriched with execution events

**Handling Preallocated Data Values.** To illustrate the automatic handling of preallocated data values, it is assumed that another user has already provided value $false$ for $\gamma^I_{Approval}$ in state $\sigma^I_{Decision}$, i.e., $\gamma^I_{Approval}.d_\gamma = Preallocated$ holds. Note that this provision of a value outside of the normal execution order is a feature of the operational semantics of lifecycle processes and not merely part of the example. As $\sigma^I_{Decision}$ is not currently the active state (i.e., $\mu_\sigma = Waiting$), decision step $\gamma^I_{Approval}$ is not evaluated. When reaching $\gamma^I_{Approval}$ from $\gamma^I_{Date}$ after a state change, $\gamma^I_{Approval}$ receives marking $\mu_\gamma = Enabled$. Instead of creating an "attribute value request" event, the combination of data marking $d_\gamma = Preallocated$ and marking $\mu_\gamma = Enabled$ immediately switches data marking to $d_\gamma = Assigned$. Consequently, as no attribute value request event has been raised beforehand, the completion event for providing a value is omitted. As $\gamma^I_{Approval}$ is a decision step, value $false$ subsequently leads to the activation of state $\sigma^I_{Rejected}$ (cf. Fig. 1), in which $\theta^I$ terminates. Note that the end state remains active despite the termination of the lifecycle process instance. In general, the operational semantics of lifecycle processes ensure that a previously provided value requires no further user interaction by default. However, users may still change the value afterwards should they wish to do so. Overall, the user may flexibly enter and alter data, and the operational semantics ensure data integrity.

**Handling Decision Steps.** Previously, decision step $\gamma^I_{Approval}$ was provided with a preallocated data value and state $\sigma^I_{Rejected}$ was reached, but the details pertaining to the handling of decision steps were omitted. In the following, the handling of a generic decision step $\gamma^I_{Dec}$ with $\gamma^I_{Dec}.P^I \neq \emptyset$ is discussed in detail.

The discussion uses the standard processing case $\gamma_{Dec}^I.\phi^I = \perp$, i.e., initially $\gamma_{Dec}^I$ has no preallocated data value. Due to the presence of one or more predicate steps $\rho^I \in \gamma_{Dec}^I.P^I$ representing decision options, more intermediate steps are necessary for the handling of decision steps when compared to ordinary steps. Until a completion event occurs after a value has been provisioned for a decision step $\gamma_{Dec}^I$, the decision step behaves identically to an ordinary step. Initially, when $\gamma_{Dec}^I$ has marking $\mu_\gamma = Enabled$ and $d_\gamma = Unassigned$, an attribute value request event is raised, a data value will be provided, and subsequently a completion event erases the "attribute value request" event from the event storage. At this point, the predicate steps $\rho^I$ of the decision step is evaluated and it is determined which decision options apply. For each predicate step $\rho^I$, its expression representing the predicate is evaluated.

For decision step $\gamma_{Approval}^I$, two predicate steps $\rho_{true}^I$ and $\rho_{false}^I$ exist. The predicate steps are equipped with expressions representing the actual predicate, $\lambda_{true} : [Approval] == true$ and $\lambda_{false} : [Approval] == false$, respectively (cf. Fig. 1). On provision of a value (w.l.o.g. it is assumed this value is $false$) for $\gamma_{Approval}^I$, each predicate step is evaluated. For $\rho_{true}^I$, this evaluation returns false and accordingly marking $\mu_\rho = Bypassed$ is set. Marking $Bypassed$ indicates that this decision option is not valid and subsequent execution paths cannot be taken. For $\rho_{false}^I$, the expression $\lambda_{false} : [Approval] == false$ evaluates to $true$ and $\mu_\rho = Activated$ is set. The markings of predicate steps $\rho_{true}^I$ and $\rho_{false}^I$ are shown in Fig. 3.

Once each predicate step $\rho^I$ has been evaluated, the results affect the marking of the decision step $\gamma_{Dec}^I$ itself. In general, two cases need to be distinguished.

First, if all predicate steps possess marking $\mu_\rho = Bypassed$, the decision step $\gamma_{Dec}^I$ must be marked as $Blocked$. This marking indicates that the provisioned value did not lead to a successful evaluation of the decision options, and the execution of the lifecycle process can therefore not proceed. To rectify the issue, a new value for $\gamma_{Dec}^I$ needs to be provided. In turn, this triggers another evaluation of the predi-

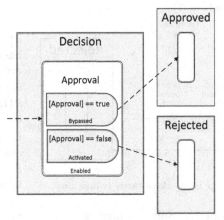

**Fig. 3.** Decision step execution status

cate steps, ensuring that process execution is not stuck when an invalid value has been provisioned.

In the second case, at least one of the predicate steps' expressions evaluate to $true$ and process execution may proceed. This is the case in Fig. 3 with the predicate steps of $\gamma_{Approval}^I$. Decision step $\gamma_{Dec}^I$ becomes marked as $\mu_\gamma = Activated$. Subsequently, a series of marking changes occurs, leading to the decision step and its predicate steps with marking $\mu_\rho = Activated$ to be marked as $Unconfirmed$.

For decision steps, this raises several challenges that need to be solved in regard to its outgoing transitions becoming enabled. First, to allow modeling sophisticated decisions, it is permitted that predicates overlap, i.e., for a given data value, two or more predicates may evaluate to *true*. In turn, this might lead to the simultaneous enabling of outgoing transitions of the predicate steps. This is not permitted, as for example two states may be become active at the same time. For this reason, lifecycle processes perform a *priority evaluation* when multiple transitions are about to become enabled. Each transition $\tau^I$ has an assigned priority $\tau^I.p$ (cf. Definition 2). Only the transition with the highest priority becomes enabled, whereas all others are marked as *Bypassed*. The priorities are assigned by the process modeler at design time, allowing for full control over decision options with overlapping predicates.

**Handling Backwards Transitions and Invalidation Events.** Consider again the example from before, where at the moment the lifecycle process has terminated and $\sigma^I_{Rejected}$ is the active state. In this situation, a user decides he wants to revise his decision for approval and thus change the value of $\gamma^I_{Approval}$ from *false* to *true*. After $\sigma^I_{Rejected}$ had become active, two backwards transition instances $\psi^I_{ToInit}$ and $\psi^I_{ToDec}$ became confirmable, i.e., their marking changed to $\mu_\psi = Confirmable$. In consequence, two *"backwards transition confirm request"* events were created, one for each backwards transition, and then were stored in $E_\theta$ (Fig. 4).

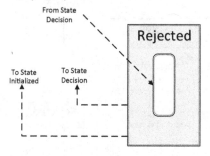

**Fig. 4.** Backwards transitions

This allows going back to state $\sigma^I_{Initialized}$, by using $\psi^I_{ToInit}$, or going back to $\sigma^I_{Decision}$, by using $\psi^I_{ToDec}$. However, only one state may be active at once. Therefore, only one backwards transition may be taken. To revise the value of $\gamma^I_{Approval}$, $\psi^I_{ToDec}$ must be confirmed. Confirming $\psi^I_{ToDec}$ causes its marking to change to $\mu_\psi = Ready$. Analogously to a step, a completion event is created, which removes the corresponding "backwards transition confirm request" event from $E_\theta$. Subsequently, $\sigma^I_{Rejected}$ is marked as $\mu_\sigma = Waiting$ and $\sigma^I_{Decision}$ is marked as $\mu_\sigma = Activated$, which allows altering the value of $\gamma^I_{Approval}$ to true. As $\sigma^I_{Rejected}$ is no longer active, $\psi^I_{ToInit}$ and $\psi^I_{ToDec}$ become marked as $\mu_\psi = Waiting$. Resetting the markings of both $\psi^I_{ToInit}, \psi^I_{ToDec}$, and $\sigma^I_{Rejected}$ to *Waiting* enables their reuse, e.g., if the value of $\gamma^I_{Approval}$ remains unchanged and the same path is taken again.

With state $\sigma^I_{Decision}$ becoming active again, it is possible to change the value of $\sigma^I_{Approval}$. However, the "backwards transition confirm request" event belonging to $\psi^I_{ToInit}$ is still stored in $E_\theta$, despite $\psi^I_{ToInit}$ having been marked with $\mu_\psi = Waiting$, i.e., confirming $\psi^I_{ToInit}$ is no longer possible. Obviously, this constitutes an inconsistency between the forms and the lifecycle process. The form displays a button with the option that $\psi^I_{ToInit}$ can be confirmed, but on pressing the button the PHILharmonicFlows system produces an error and other, possibly worse, side effects. As a consequence, the operational semantics include invalidation events, with the purpose to remove invalid or obsolete execution events from event storage $E_\theta$. An invalidation event occurs when entities with a request event, e.g., backwards transitions, are not successfully completed, but become changed due to other circumstances, e.g., the confirmation of another backwards transition.

Request events, completion events, and invalidation events are used in many more situations than discussed above. The basic principles, however, are always the same, and, embedded in the overall operational semantics, provide a robust and flexible way to acquire data values for lifecycle processes. The imperative-like modeling style of lifecycle processes, from which forms can be auto-generated directly, significantly reduces modeling time and efforts. The operational semantics provide the necessary flexibility to users interacting with the forms. Furthermore, the use of forms and the emulation of standard form behavior simplifies the usage of the PHILharmonicFlows system for non-expert users.

Overall, this section described the functional aspects of the operational semantics of lifecycle processes. The technical implementation of these operational semantics with the *Process Rule Framework* is presented in Sect. 4.

# 4   The Process Rule Framework

In the description of the operational semantics of lifecycle processes (cf. Sect. 3), at the lowest level, progress is driven by the change of markings. Marking changes elicit the creation of execution events, which, in turn, results in user actions, e.g., the provision of a data value for an attribute. This user interaction is reflected in the lifecycle process by setting new markings. This may be viewed as a chain of events, and, consequently, event-condition-action rules are used as the technical basis for the technical implementation of the operational semantics. In PHILharmonicFlows, a specialized variant of ECA rules, denoted as *process rules,* is employed for this purpose. Process rules and the means to specify them constitute one part of the *process rule framework.* To create an execution sequence, such as the one described in Sect. 3.2, process rules need to form *process rule cascades,* i.e., a rule triggers an event, which may trigger another rule, which again triggers an event. Furthermore, process rules are uniquely suited to deal with the different eventualities emerging during the execution of lifecycle processes. For example, a state $\sigma^I$ may become active in context of normal process execution progress or due to the use of a backwards transition $\psi^I$. Subsequently, different follow-up measures may be required, e.g., the resetting of markings for steps $\gamma^I \in \sigma^I.\Gamma^I$ in case the backwards transition became activated.

The basic definition of a process rule is given in Definition 3. In order to distinguish these symbols from symbols used in the definition of object instances, superscript $^R$ is used.

**Definition 3.** *A process rule $p^R$ has the form $(\epsilon, e^T, C^R, A^R)$ where*

- $\epsilon$ *is an event triggering the evaluation of the rule.*
- $e^T$ *is an entity type, e.g., a step type $\gamma^T$.*
- $C^R$ *is a set of preconditions in regard to $e^T$.*
- $A^R$ *is a set of effects.*

Process rules $p^R$ may be evaluated, i.e., their preconditions $C^R$ are checked and, if all are fulfilled, the effects $A^R$ are applied. An evaluation is triggered when the event $\epsilon$ occurs. Events $\epsilon$ are always raised by a particular entity instance $e^I$, e.g., a step $\gamma^I$ or a transition $\tau^I$. $e^T$ is an entity type that provides the context for defining conditions and effects. Furthermore, it provides an implicit precondition, meaning a rule is not evaluated if the entity instance $e^I$ raising $\epsilon$ was not created from $e^T$. Preconditions $C^R$ check different properties of an entity, e.g., whether the entity has a specific marking. Effects $A^R$ apply different effects to an entity, e.g., setting the marking of an entity. Note that preconditions and effects are not limited to properties belonging to instances of $e^T$. They may also access or set properties of neighbor entities. For example, a rule defined for a step $\gamma^T$ may have effects that set markings for the outgoing transitions $\tau^I_{out} \in \gamma^I.T^I_{out}$ of the corresponding step instance.

```
public class MarkingRuleMr14A : AbstractEffectRule<TransitionInstance>
{
    public MarkingRuleMr14A()
    {
        Name = "Marking Rule Mr14A";
        ShortDescription = "Marking steps and predicate steps as Enabled if the transition is internal and marked as Ready";

        PreconditionFor< TransitionInstance,TransitionMarkings>(trans => trans.Marking).IsMarked(TransitionMarkings.Ready);
        PreconditionFor< TransitionInstance,bool>(trans => trans.IsExternal).SatisfiesPredicate(x => !x);

        //If the step is a decision step, mark its predicate steps also as Enabled
        EffectForEach<PredicateStepInstance, StepMarkings>(
            trans => trans.Target.Cast<DecisionStepInstance>().PredicateSteps.Select(x => x.Marking))
            .AssignMarking(StepMarkings.Enabled)
            .When(trans => trans.Target is DecisionStepInstance);

        //Mark the target micro step as Enabled
        EffectFor<AbstractStepInstance, StepMarkings>(trans => trans.Target.Marking).AssignMarking(StepMarkings.Enabled);
    }
}
```

**Fig. 5.** Fluent interface definition of a marking rule in code

In the PHILharmonicFlows implementation, process rules are created using a domain-specific language. Figure 5 shows an example of how a process rule is represented. Process rules are often subject to change, as new features for PHILharmonicFlows are added or errors in lifecycle process execution are resolved. In order to be able to quickly adapt a process rule, the process rule framework uses a *fluent interface* for process rule specification, i.e., the domain-specific language is structured to resemble natural prose text. This allows for both a high readability and maintainability.

The operational semantics introduced in Sect. 3 allow identifying different use cases for process rules. For example, one type of process rule raises execution events based on specific markings, while another type reacts to user input and sets appropriate markings. Accordingly, process rules are subdivided based on their purpose. The type determines the general type of preconditions and effects, e.g., preconditions of *marking rules* check predominantly for specific markings. The different types of process rules are summarized in Table 1. *Request rule*, *completion rule*, and *invalidation rule* are subsumed under the term *execution rule* (ER).

**Table 1.** Overview over the types of process rules

| Rule | Abbreviation | Event | Preconditions | Effects |
|------|-------------|-------|---------------|---------|
| Marking rule | MR | Marking event | Markings | Markings |
| Request rule | QR | Marking event | Markings | Request event |
| Completion rule | CR | Marking event | Markings | Completion event |
| Invalidation rule | IR | Marking event | Markings | Invalidation event |
| Reaction rule | RR | User input event | User input | Markings |

The most common event that is raised during the execution of a lifecycle process instance is a *marking event*. An entity instance $e^I$ raises a marking event whenever its marking $e^I.\mu$ is changed. In order to determine which process rule needs to be applied, the event is gathered by the *process rule manager* (PRM) of the lifecycle process. The process rule manager is a small and lightweight execution engine for process rules and constitutes the other part of the process rule framework. Figure 6 shows a schematic view of the process rule manager and its interactions with the lifecycle process and the (auto-generated) forms.

Starting at ① in Fig. 6, data has been entered into a form field. The data is then passed on to the lifecycle process $\theta^I$ and the corresponding step $\gamma^I$. As $\gamma^I$ has received a value, the step raises a user input event ②. The event is passed on to the process rule manager, which receives all events from its corresponding lifecycle process $\theta^I$ and evaluates appropriate rules, i.e., process rules $p^R$ with $p^R.e^T = \sigma^T$ are not evaluated if the entity creating the event has type $\gamma^T$. Note that this implicit precondition significantly reduces the search space for process rule application. Once the PRM has identified all currently applicable rules, the effects of each rule are applied. In the example, the PRM identifies a reaction rule and applies its effects to the appropriate entities in the lifecycle process ③.

Applying the effects from the reaction rule application raises marking events, which trigger a completion rule and a marking rule in the PRM. The completion rule raises a completion event ④, removing the request event for the mandatory form field from event storage $E_\theta$ of $\theta^I$. In parallel, the marking rule sets markings for the outgoing transitions $T^I$ of step $\gamma^I$. This again creates marking events, resulting in a cascade of marking rules, i.e., the PRM alternates between ② and ③ in Fig. 6. The process rule cascade stops when the next step becomes marked with $\mu_\gamma = Enabled$. This raises a request event, which is deposited in event storage $E_\theta$ ④. When a user views a form, the updated event storage $E_\theta$ and the

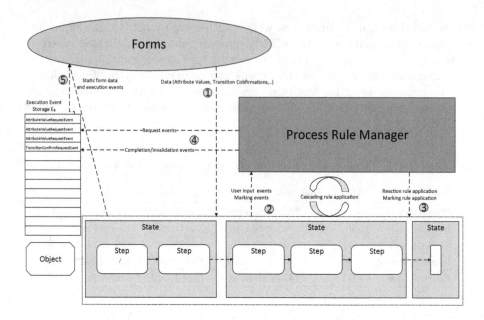

**Fig. 6.** Process rule manager and schematic process rule application

static form data are combined into a new form ⑤. When the user enters data for the next form field, the cycle starts again at ①.

When a user fills out a form, the form is expected to tell the user immediately which form field is required next after providing data for a form field. Long processing times are prohibitive for the usability of the PHILharmonicFlows process management system. In order to have full control over processing times and the tight connection of process rules with lifecycle process entities, it was decided to implement the PRM as a custom, lightweight rule engine. A custom PRM implementation offers a fine-grained control over process rule application. By default, the PRM handles events in the order in which they arrive (FIFO principle). However, in several cases, the handling of specific events needed to be delayed or accelerated in order to ensure a form processing in compliance with the operational semantics. For example, an event $e_\tau$ triggering the transition $\tau^I$ from a source state $\sigma^I_{source}$ to a target state $\sigma^I_{target}$ is, under certain circumstances, raised before all steps $\gamma^I \in \sigma^I_{source}.\Gamma^I$ have been processed. This results in errors in the application of the process rules, as the target state $\sigma^I_{target}$ already received $\mu_\sigma = Activated$ when events from $\gamma^I \in \sigma^I_{source}.\Gamma^I$ arrive at the PRM. To prevent such errors, the handling of the state transition event $e_\tau$ must be delayed until all steps $\gamma^I$ in the source state $\sigma^I_{source}$ have finished processing. In consequence, the PRM was extended with a priority queue that retains the FIFO principle, but allows assigning different priorities to events, accelerating or delaying them as needed.

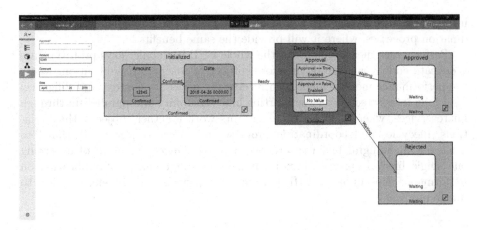

**Fig. 7.** Run-time environment of PHILharmonicFlows, executing a Transfer lifecycle process

Figure 7 shows the run-time environment of the PHILharmonicFlows prototype, which is currently executing a *Transfer* object. Besides the advantages for the application of process rules, the lightweight nature of the PRM also proves beneficial for the transition of PHILharmonicFlows to a microservice-based architecture. The PRM was initially conceived as a monolithic rule engine, i.e., all lifecycle processes use the same instance of the PRM. Currently, PHILharmonicFlows is moving towards a hyperscale architecture [2], based on a microservice framework. A microservice is a lightweight and independent service that performs single functions and interacts with other microservices in order to realize a software application. In this new hyperscale architecture, an object and its lifecycle process are implemented as a single microservice. A continued use of a single PRM instance generates a significant performance overhead due to the necessary message exchanges between the PRM and the microservices. The single PRM instance is a bottleneck and puts a limit on the scalability of the microservice-based architecture, i.e., it would no longer be warranted to designate the PHILharmonicFlows system as hyperscale. Furthermore, the communication overhead and the delays of process rule application in the PRM, due to the high number of events simultaneously created by the object instance microservices, would negatively affect the performance of the auto-generated forms.

Fortunately, the lightweight nature of the PRM offers a satisfactory solution. By integrating an instance of the PRM into the microservice of each object instance, no message exchanges between PRM and lifecycle process are required. Furthermore, a PRM instance is only responsible for exactly one lifecycle process instance. This eliminates the delays in rule application due to the processing of other lifecycle processes. This solution offers sufficient performance for displaying dynamic forms while retaining the hyperscale property of the PHILharmonicFlows microservice-based architecture. The approach to integrate a PRM instance into a microservice will also be used with the implementation of coordination processes, where it will provide the same benefits.

Performance measurements of the whole PHILharmonicFlows prototype are a delicate endeavor and are therefore subject to a separate publication, where the performance measurements can receive the necessary context and diligence. As a fully integrated system supporting high scalability and parallelism through microservices, where also multiple concepts work together (objects, their relations, lifecycles and coordination processes to govern object interactions) to achieve a meaningful business process, a performance evaluation of executing one single lifecycle process is not particularly enlightening. A publication on performance aspects of the PHILharmonicFlows systems is therefore subject to future publications.

## 5   Related Work

Opus [8,10] is a data-centric process management system that bases its processes on Petri nets. Petri nets are a popular and well-established formalism for modeling business processes. Additionally, Petri nets provide several verification techniques, e.g., soundness checks or deadlock detection, which may also be applied to verify process model correctness. In Opus, the Petri net formalism is extended with structured data tuples, which substitute the places of a standard Petri net. The transitions of this extended Petri net provide operations on the data, e.g., operations derived from operations of relational algebra. The Opus approach does not support automatically generating forms from process models. Furthermore, Petri nets are inherently more rigid in their execution and do not provide the same built-in flexibility as PHILharmonicFlows and the operational semantics of lifecycle processes. However, Opus is capable to explicitly model the different execution paths to provide flexible process execution. Opus provides an implemented prototype of the approach [9].

Case Handling [7,21,24] defines a case in terms of activities and data objects. Activities are ordered in an acyclic graph in which edges represent precedence relations. To execute an activity, all precedence relations before the activity must be fulfilled. Furthermore, the execution of an activity is restricted by data bindings. A data binding represents a condition so that a data object must have a specific value at run-time. The values of the data objects are acquired by forms, which are associated with activities. While case handling possesses forms, it is unclear whether these can be auto-generated from the activities or must be created manually. While both case handling and PHILharmonicFlows use an acyclic graph to represent processes, the operational semantics for lifecycle processes in PHILharmonicFlows allows for data to be acquired at any point in time. A case acquires data by activities and that activities have a precedence relation, the same flexibility in regard to data acquisition is not possible. A detailed comparison between case handling and object-aware process management was performed in [4].

The Guard-Stage-Milestone (GSM) meta-model [14] is a declarative notation for specifying artifact-centric processes [5,13,17]. An artifact consists of an information model, i.e., attributes and a lifecycle model. The lifecycle model is specified using GSM. Its operational semantics are based on Precedent-Antecedent-Consequent rules and possess different, but semantically equivalent formulations [6]. In GSM, tasks provide the means to write attributes and acquire data. Because of being declarative, guards, stages and milestones may be used in such a way that flexible data acquisition, within certain constraints, becomes possible. Tasks may be defined so that attributes may be written at any point in time and may be restricted, if necessary. Lifecycle processes defined in GSM are able to react to the newly acquired data and might be more flexible than lifecycle processes in PHILharmonicFlows. However, as a severe drawback, much of this flexibility in data acquisition must be implemented by the process modeler. Furthermore, the is no auto-generation of forms from GSM-specified lifecycle models within the artifact-centric approach.

CMMN [18] is a standard notation for case management as proposed by OMG. The notation is closely inspired by GSM and its execution semantics and therefore inherits many of the same advantages and disadvantages. As such, flexibility in practice has to be provided by the model and is not simply provided by the operational semantics. Also, automatic generation of dynamic forms it not supported.

Fragment-based case management [3,12] is a promising approach that defines business processes in form of pre-specified process fragments. Fragments are specified using activities and control flow. The execution order of fragments is, in principle, completely free, i.e., any process process fragment may be executed at any point in time. This freedom is only limited by data conditions that govern the activation of a process fragment, i.e., a process fragment may only be executed if the data conditions are met. In turn, process fragments may generate new data to fulfill other data conditions and subsequently enable more process fragments. As data is mostly required to enable process fragments and their activities, it is unclear whether automatic form generation with dynamic control by the process is achievable. Through breaking rigid control flow ordering of activities with the use of process fragments, their flexible execution may only be achieved by modeling appropriate data conditions and is not automatically provided by the operational semantics, as accomplished in PHILharmonicFlows.

## 6  Summary and Outlook

The PHILharmonicFlows project is a full, though prototypical, data-centric process management system incorporating modeling and execution environments. One aspect of this system is to have highly flexible executions of object lifecycle processes that require minimal effort on part of the process modeler. The scientific contribution of this paper is to show *that* the intended level of flexibility has been achieved. As proof, it is shown exactly *how* the flexibility is achieved by describing its implementation and inner workings in full detail.

The technical implementation of the operational semantics of lifecycle processes in object-aware process management is achieved by *process rules*, which govern the changing of markings and the creation of execution events. This paper presented the process rule framework, for which two aspects need to be emphasized. First, the process rule framework ensures that lifecycle processes execute correctly and also provides the technical basis for the operational semantics of coordination processes in PHILharmonicFlows. Coordination processes, as the name suggests, coordinate lifecycle processes of multiple objects, so that complex business processes can be realized. Its operational semantics will be based on the process rule framework as well. Second, a performant, efficient and lightweight technical basis for enacting lifecycle processes and coordination processes is crucial for the transition of PHILharmonicFlows to a hyperscale architecture. The operational semantics of lifecycle processes provide a flexible acquisition of data, while modeling efforts are minimal due to an modeling style that is akin to an imperative style. The flexibility is not provided by the lifecycle process model, but by the operational semantics. The model of the lifecycle process and the operational semantics together provide the means to auto-generate dynamic forms.

**Acknowledgments.** This work is part of the ZAFH Intralogistik, funded by the European Regional Development Fund and the Ministry of Science, Research and the Arts of Baden-Württemberg, Germany (F.No. 32-7545.24-17/3/1)

# References

1. Andrews, K., Steinau, S., Reichert, M.: Enabling fine-grained access control in flexible distributed object-aware process management systems. In: 21st IEEE International Conferences on Enterprise Distributed Object Computing (EDOC) (2017)
2. Andrews, K., Steinau, S., Reichert, M.: Towards hyperscale process management. In: 8th International Workshop on Enterprise Modeling and Information Systems Architectures (EMISA), CEUR Workshop Proceedings, pp. 148–152. CEUR-WS.org (2017)
3. Beck, H., Hewelt, M., Pufahl, L.: Extending fragment-based case management with state variables. In: Dumas, M., Fantinato, M. (eds.) BPM 2016. LNBIP, vol. 281, pp. 227–238. Springer, Cham (2017). https://doi.org/10.1007/978-3-319-58457-7_17
4. Chiao, C.M., Künzle, V., Reichert, M.: Enhancing the case handling paradigm to support object-aware processes. In: 3rd International Symposium on Data-Driven Process Discovery and Analysis (SIMPDA), CEUR Workshop Proceedings, pp. 89–103. CEUR-WS.org (2013)
5. Cohn, D., Hull, R.: Business artifacts: a data-centric approach to modeling business operations and processes. Bull. IEEE Comput. Soc. Tech. Committee Data Eng. **32**(3), 3–9 (2009)
6. Damaggio, E., Hull, R., Vaculín, R.: On the equivalence of incremental and fixpoint semantics for business artifacts with Guard-Stage-Milestone lifecycles. Inf. Syst. **38**(4), 561–584 (2013)

7. Guenther, C.W., Reichert, M., van der Aalst, W.M.P.: Supporting flexible processes with adaptive workflow and case handling. In: IEEE 17th Workshop on Enabling Technologies: Infrastructure for Collaborative Enterprises, pp. 229–234 (2008)
8. Haddar, N., Tmar, M., Gargouri, F.: A framework for data-driven workflow management: modeling, verification and execution. In: Decker, H., Lhotská, L., Link, S., Basl, J., Tjoa, A.M. (eds.) DEXA 2013. LNCS, vol. 8055, pp. 239–253. Springer, Heidelberg (2013). https://doi.org/10.1007/978-3-642-40285-2_21
9. Haddar, N., Tmar, M., Gargouri, F.: Opus framework: a proof-of-concept implementation. In: IEEE/ACIS 14th International Conference on Computer and Information Science (ICIS), pp. 639–641 (2015)
10. Haddar, N., Tmar, M., Gargouri, F.: A data-centric approach to manage business processes. Computing 98(4), 375–406 (2016)
11. Haisjackl, C., et al.: Understanding declare models: strategies, pitfalls, empirical results. Softw. Syst. Model. 15(2), 325–352 (2016)
12. Hewelt, M., Weske, M.: A hybrid approach for flexible case modeling and execution. In: La Rosa, M., Loos, P., Pastor, O. (eds.) BPM 2016. LNBIP, vol. 260, pp. 38–54. Springer, Cham (2016). https://doi.org/10.1007/978-3-319-45468-9_3
13. Hull, R., et al.: Business artifacts with Guard-Stage-Milestone lifecycles: managing artifact interactions with conditions and events. In: 5th ACM International Conference on Distributed Event-based System (DEBS), pp. 51–62. ACM (2011)
14. Hull, R., et al.: Introducing the Guard-Stage-Milestone approach for specifying business entity lifecycles. In: Bravetti, M., Bultan, T. (eds.) WS-FM 2010. LNCS, vol. 6551, pp. 1–24. Springer, Heidelberg (2011). https://doi.org/10.1007/978-3-642-19589-1_1
15. Künzle, V., Reichert, M.: A modeling paradigm for integrating processes and data at the micro level. In: Halpin, T., et al. (eds.) BPMDS/EMMSAD-2011. LNBIP, vol. 81, pp. 201–215. Springer, Heidelberg (2011). https://doi.org/10.1007/978-3-642-21759-3_15
16. Künzle, V., Reichert, M.: PHILharmonicFlows: towards a framework for object-aware process management. J. Softw. Maint. Evol.: Res. Pract. 23(4), 205–244 (2011)
17. Nigam, A., Caswell, N.S.: Business artifacts: an approach to operational specification. IBM Syst. J. 42(3), 428–445 (2003)
18. Object Management Group: Case Management Model and Notation (CMMN), Version 1.1 (2016)
19. Pesic, M., Schonenberg, H., van der Aalst, W.M.P.: DECLARE: full support for loosely-structured processes. In: 11th IEEE International Conference on Enterprise Distributed Object Computing (EDOC), p. 287 (2007)
20. Pichler, P., Weber, B., Zugal, S., Pinggera, J., Mendling, J., Reijers, H.A.: Imperative versus declarative process modeling languages: an empirical investigation. In: Daniel, F., Barkaoui, K., Dustdar, S. (eds.) BPM 2011. LNBIP, vol. 99, pp. 383–394. Springer, Heidelberg (2012). https://doi.org/10.1007/978-3-642-28108-2_37
21. Reijers, H.A., Rigter, J.H.M., van der Aalst, W.M.P.: The case handling case. Int. J. Coop. Inf. Syst. 12(03), 365–391 (2003)
22. Steinau, S., Andrews, K., Reichert, M.: The relational process structure. In: Krogstie, J., Reijers, H.A. (eds.) CAiSE 2018. LNCS, vol. 10816, pp. 53–67. Springer, Cham (2018). https://doi.org/10.1007/978-3-319-91563-0_4
23. Steinau, S., Künzle, V., Andrews, K., Reichert, M.: Coordinating business processes using semantic relationships. In: 19th IEEE Conference on Business Informatics (CBI), pp. 33–43. IEEE Computer Society Press (2017)

24. van der Aalst, W.M.P., Weske, M., Grünbauer, D.: Case handling: a new paradigm for business process support. Data Knowl. Eng. **53**(2), 129–162 (2005)
25. Weber, B., Mutschler, B., Reichert, M.: Investigating the effort of using business process management technology: results from a controlled experiment. Sci. Comput. Program. **75**(5), 292–310 (2010)

# Towards Semantic Process Mining Through Knowledge-Based Trace Abstraction

G. Leonardi[1], M. Striani[2], S. Quaglini[3], A. Cavallini[4], and S. Montani[1(✉)]

[1] DISIT, Computer Science Institute, Università del Piemonte Orientale,
Alessandria, Italy
stefania.montani@uniupo.it
[2] Department of Computer Science, Università di Torino, Turin, Italy
[3] Department of Electrical, Computer and Biomedical Engineering,
Università di Pavia, Pavia, Italy
[4] I.R.C.C.S. Fondazione "C. Mondino" - on behalf of the Stroke Unit Network (SUN)
Collaborating Centers, Pavia, Italy

**Abstract.** Many information systems nowadays record data about the process instances executed at the organization in the form of *traces* in a log. In this paper we present a framework able to convert actions found in the traces into higher level concepts, on the basis of domain knowledge. Abstracted traces are then provided as an input to semantic process mining.

The approach has been tested in the medical domain of stroke care, where we show how the abstraction mechanism allows the user to mine process models that are easier to interpret, since unnecessary details are hidden, but key behaviors are clearly visible.

**Keywords:** Semantic process mining ·
Knowledge-based trace abstraction · Medical applications

## 1 Introduction

Most commercial information systems, including those adopted by many health care organizations, record information about the executed process instances in a log [29]. The log stores the sequences (*traces* [29] henceforth) of actions that have been executed at the organization, typically together with key execution parameters, such as times, cost and resources. Logs can be provided in input to **process mining** [29,30] algorithms, a family of a-posteriori analysis techniques able to extract non-trivial knowledge from these historic data; within process mining, *process model discovery* algorithms, in particular, take as input the log traces and build a process model, focusing on its control flow constructs. Classical process mining algorithms, however, provide a purely syntactical analysis, where actions in the traces are processed only referring to their names. Action names

© IFIP International Federation for Information Processing 2019
Published by Springer Nature Switzerland AG 2019
P. Ceravolo et al. (Eds.): SIMPDA 2017, LNBIP 340, pp. 45–64, 2019.
https://doi.org/10.1007/978-3-030-11638-5_3

are strings without any semantics, so that identical actions, labeled by synonyms, will be considered as different, or actions that are special cases of other actions will be processed as unrelated.

Relating *semantic structures*, such as ontologies, to actions in the log, not only can solve the synonyms issue, but also can enable trace comparison and process mining techniques to work at *different levels of abstraction* (i.e., at the level of instances and/or concepts) and, therefore, to mask irrelevant details, to promote reuse, and, in general, to make process analysis much more flexible and reliable.

In fact, it has been observed that human readers are limited in their cognitive capabilities to make sense of large and complex process models [1,33], while it would be often sufficient to gain a quick overview of the process, in order to familiarize with it in a short amount of time. Of course, deeper investigations can still be conducted, subsequently, on the detailed (ground) process model.

Interestingly, **semantic process mining**, defined as the integration of semantic processing capabilities into classical process mining techniques, has been recently proposed in the literature (see Sect. 5). However, while more work has been done in the field of semantic *conformance checking* (another branch of process mining) [10,13], to the best of our knowledge semantic *process model discovery* needs to be further investigated.

In this paper, we present a **knowledge-based abstraction mechanism** (see Sect. 2), able to operate on log traces. In our approach:

- actions in the log are mapped to the ground terms of an *ontology*;
- a *rule base* is exploited, in order to identify which of the multiple ancestors of an action should be considered for abstracting the action itself. *Medical knowledge and contextual information* are resorted to in this step;
- when a set of consecutive actions on the trace abstract as the same ancestor, they are merged into the same abstracted *macro-action*, labeled as the common ancestor at hand. This step requires a proper treatment of delays and/or actions in-between that descend from a different ancestor.

Our abstraction mechanism is then provided as an input to **semantic process mining** (see Sect. 3). In particular, we rely on classical *process model discovery* algorithms embedded in the open source framework ProM [32], made semantic by the exploitation of domain knowledge in the abstraction phase.

We also describe our experimental work (see Sect. 4) in the field of stroke care, where the application of the abstraction mechanism on log traces has allowed us to mine simpler and more understandable (from the clinical point of view) process models.

## 2   Knowledge-Based Trace Abstraction

In our framework, trace abstraction has been realized as a multi-step mechanism. The following subsections describe the various steps.

## 2.1   Ontology Mapping

As a first step, every action in the trace to be abstracted is mapped to a ground term of an **ontology**, formalized resorting to domain knowledge.

In our current implementation, we have defined an ontology related to the field of stroke management, where ground terms are patient management actions, while abstracted terms represent medical goals. Figure 1 shows an excerpt of the stroke domain ontology, formalized through the Protègè editor.

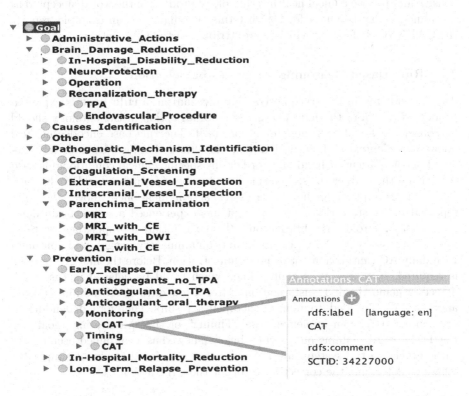

**Fig. 1.** An excerpt from the stroke domain ontology. The annotation shown for CAT (computer assisted tomography) reports its SNOMED code

In particular, a set of classes, representing the main goals in stroke management, have been identified, namely: "Administrative Actions" (such as hospital admission and discharge), "Brain Damage Reduction","Causes Identification", "Pathogenetic Mechanism Identification", "Prevention", and "Other". These main goals can be further specialized into subclasses, according to more specific goals (e.g., "Parenchima Examination" is a subgoal of "Pathogenetic Mechanism Identification", while "Early Relapse Prevention" is a subgoal of "Prevention"), down to the ground actions, that will implement the goal itself.

Some actions in the ontology can be performed to implement different goals. For instance, a Computer Assisted Tomography (CAT) can be used to monitor therapy efficacy, or to assess therapy starting time (see class "Prevention" in Fig. 1). The proper goal to be used in the abstraction phase will be selected on the basis of the context of execution, as formalized in the rule base, described in the following subsection.

All ground actions are being mapped to SNOMED concepts[1]. The ontology partially showed in Fig. 1 is therefore being integrated with the most comprehensive and precise clinical health terminology product in the world, accepted as a common global language for health terms. The figure, as an example, reports the CAT SNOMED code, which is an attribute of the CAT action.

## 2.2  Rule-Based Reasoning for Ancestor Selection

As a second step in the trace abstraction mechanism, a **rule base** is exploited to identify which of the multiple ancestors of an action in the ontology should be considered for abstracting the action itself. The rule base encodes medical knowledge. Contextual information (i.e., the actions that have been already executed on the patient at hand, and/or her/his specific clinical conditions) is used to activate the correct rules. The rule base has been formalized in Drools [20].

As an example, referring to the CAT action mentioned above, the rules reported below state that, if the patient has experienced a severe brain damage and suffers from atrial fibrillation, s/he must initiate a proper therapy. Such a therapy starts with ASA (a class of anti-inflammatory drugs), and continues with daily AC (anti-coagulant drug) administration. Before the first AC, a CAT is required, to assess AC starting time, which could be delayed in case CAT detects a hemorrhagic transformation. After a few days of AC administration, another CAT is needed, to monitor therapeutic results. Therefore, depending on the context, CAT can implement the "Timing" or the "Monitoring" goal (see Fig. 1). Forward chaining on the rules below (showed as a simplified pseudocode with respect to the system internal representation, for the sake of simplicity) allows to determine the correct ancestor for the CAT action.

```
rule "SevereDamage"
    when
        (
            Damage(value > threshold) &&
            AtrialFibrillation(value=true)
        )
    then
        logicalInsertFact (DamFib);
end

rule "Fibrillation1"
    when
```

---

[1] http://www.snomed.org/snomed-ct, last accessed on 02/08/2018.

```
        existInLogical(DamFib) &&
        isBefore("CAT", "AC")
    then
        setAncestorName("CAT", "Timing");
end

rule "Fibrillation2"
    when
        existInLogical(DamFib) &&
        isAfter("CAT", "AC")
    then
        setAncestorName("CAT", "Monitoring");
end
```

## 2.3    Trace Abstraction

Once the correct ancestor of every action has been identified, trace abstraction can be completed.

In this last step, when a set of consecutive actions on the trace abstract as the same ancestor, they have to be merged into the same abstracted **macro-action**, labeled as the common ancestor at hand. This procedure requires a proper treatment of *delays*, and of actions in-between that descend from a different ancestor (*interleaved actions* henceforth).

Trace abstraction has been realized by means of the procedure described in Algorithm 1 below.

The function *abstraction* takes in input a log *trace*, the domain ontology *onto*, and the *level* in the ontology chosen for the abstraction (e.g., *level* = 1 corresponds to the choice of abstracting the actions up to the sons of the ontology root). It also takes in input three thresholds (*delay_th*, *n_inter_th* and *inter_th*). These threshold values have to be set by the domain expert in order to limit the total admissible delay time within a macro-action, the total number of interleaved actions, and the total duration of interleaved actions, respectively. In fact, it would be hard to justify that two ground actions share the same goal (and can thus be abstracted to the same macro-action), if they are separated by very long delays, or if they are interleaved by many/long different ground actions, meant to fulfill different goals (where the term "long" may correspond to different quantitative values in different application domains).

The function outputs an abstracted trace.

For every action $i$ in *trace*, an iteration is executed (lines 3–27). First, a macro-action $m_i$, initially containing just $i$, and sharing its starting and ending times, is created. $m_i$ is labeled referring to the ancestor of $i$ (the one identified by the rule based reasoning procedure) at the abstraction *level* provided as an input. Accumulators for this macro-action (*total-delay*, *num-inter* and *total-inter*, commented below) are initialized to 0 (lines 4–10). Then, a nested cycle is executed (lines 11–25): it considers every element $j$ following $i$ in the trace, where a trace element can be an action, or a delay between a pair of consecutive actions. Different scenarios can occur:

**Algorithm 1.** Multi-level abstraction algorithm

```
 1  abs_trace = abstraction(trace, onto, level, delay_th, n_inter_th, inter_th);
 2  abs_trace = ∅;
 3  for every i ∈ actions in trace do
 4      if (i.startFlag = yes) then
 5          create : m_i as ancestor(i, level);
 6          m_i.start = i.start;
 7          m_i.end = i.end;
 8          total_delay = 0;
 9          num_inter = 0;
10          total_inter = 0;
11          for (every j ∈ elements in trace) do
12              if (j is a delay) then
13                  total_delay = total_delay + j.length;
14              else
15                  if (ancestor(j, level)=ancestor(i, level)) then
16                      if (total_delay < delay_th ∧ num_inter <
                            n_inter_th ∧ total_inter < inter_th) then
17                          m_i.end = max(m_i.end, j.end);
18                          j.startFlag = no;
19                      end
20                  else
21                      num_inter = num_inter + 1;
22                      total_inter = total_inter + j.length;
23                  end
24              end
25          end
26      append m_i to abs_trace;
27  end
28  return abs_trace;
```

- if $j$ is a delay, *total-delay* is updated by summing the length of $j$ (lines 12–14).
- if $j$ is an action, and $j$ shares the same ancestor of $i$ at the input abstraction *level*, then $j$ is incorporated into the macro-action $m_i$. This operation is always performed, provided that *total-delay*, *number-inter* and *total-inter* do not exceed the threshold passed as an input (lines 15–19). $j$ is then removed from the actions in *trace* that could start a new macro-action, since it has already been incorporated into an existing one (line 18). This kind of situation is described in Fig. 2 (a).
- if $j$ is an action, but does not share the same ancestor of $i$, then it is treated as an interleaved action. In this case, *num-inter* is increased by 1, and *total-inter* is updated by summing the length of $j$ (lines 20–23). This situation, in the end, may generate different types of temporal constraints between macro-actions, as the ones described in Fig. 2(b) (Allen's *during* [2]) and Fig. 2(c) (Allen's *overlaps* [2]).

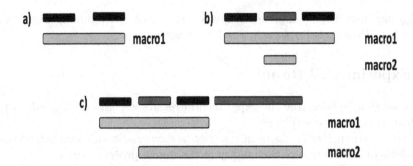

**Fig. 2.** Different trace abstraction situations: (a) two actions are abstracted to a single macro-action *macro*1, with a delay in between; (b) two actions are abstracted to a macro-action *macro*1, with an interleaved action in between, resulting in a different macro-action *macro*2 during *macro*1; (c) two actions are abstracted to a macro-action *macro*1, with an interleaved action in between, which is later aggregated to a fourth action, resulting in a macro-action *macro*2 overlapping *macro*1.

Finally, the macro-action $m_i$ is appended to *abs_trace*, that, in the end, will contain the list of all the macro-actions that have been created by the procedure (line 26).

**Complexity.** The cost of abstracting a trace is $O(actions * elements)$, where *actions* is the number of actions in the input trace, and *elements* is the number of elements (i.e., actions + delay intervals) in the input trace.

## 3   Semantic Process Mining

In our approach, process mining, made semantic by the exploitation of the abstraction mechanism illustrated above, is implemented resorting to the well-known process mining tool ProM, extensively described in [32]. ProM (and specifically its newest version ProM 6) is a platform-independent open source framework that supports a wide variety of process mining and data mining techniques, and can be extended by adding new functionalities in the form of plug-ins.

For the work described in this paper, we have exploited ProM's Heuristic Miner [35]. Heuristic Miner is a plug-in for process model discovery, able to mine process models from logs. It receives in input the log, and considers the order of the actions within every single trace. It can mine the presence of short-distance and long-distance dependencies (i.e., direct or indirect sequence of actions), with a certain degree of reliability. The output of the mining process is provided as a graph, known as the "dependency graph", where nodes represent actions, and edges represent control flow information. The output can be converted into other formalisms as well.

Currently, we have chosen to rely on Heuristics Miner, because it is known to be tolerant to noise, a problem that may affect medical logs (e.g., sometimes the logging may be incomplete). Anyway, testing of other mining algorithms available in ProM 6 is foreseen in our future work.

# 4   Experimental Results

In this section, we describe the experimental results we have conducted, in the application domain of stroke care.

First, we will present a case study, meant to showcase in an immediate fashion the effects of abstraction of the quality of the mined process model.

Then, we will discuss a more complete validation work, able to properly quantify the abstraction effects themselves.

In both studies the available log was composed of more than 15000 traces, collected at the Stroke Unit Network (SUN) collaborating centers of the Lombardia region, Italy. The number of traces varied from 266 to 1149. Traces were composed of 13 actions on average.

## 4.1   Case Study

In the case study, we wanted to test whether our capability to abstract the log traces on the basis of their semantic goals allowed to obtained process models where unnecessary details are hidden, but key behaviors are clear. Indeed, if this hypothesis holds, in our application domain it becomes easier to compare process models of different stroke units (SUs), highlighting the presence/absence of common paths, regardless of minor action changes (e.g., different ground actions that share the same goal) or irrelevant different action ordering or interleaving (e.g., sets of ground actions, all sharing a common goal, that could be executed in any order)[2].

Figure 3 compares the process models of two different SUs (SU-A and SU-B), mined by resorting to Heuristic Miner, operating on ground traces. Figure 4, on the other hand, compares the process models of the same SUs as Fig. 3, again mined by resorting to Heuristic Miner, but operating on traces abstracted according to the goals of the ontology in Fig. 1. In particular, abstraction was conducted up to level 2 in the ontology (where level 0 is the root, i.e.. "Goal").

Generally speaking, a visual inspection of the two graphs in Fig. 3 is very difficult. Indeed, these two ground processes are "spaghetti-like" [29], and the extremely large number of nodes and edges makes it hard to identify commonalities in the two models.

The abstract models in Fig. 4, on the other hand, are much more compact, and it is possible for a medical expert to analyze them.

In particular, the two graphs in Fig. 4 are not identical, but in both of them it is easy to a identify the macro-actions which corresponds to the treatment of a typical stroke patient.

---

[2] It is however worth noting that, within our framework, it is still possible to mine the process models from ground traces, and investigate them in detail as a further analysis step, if needed.

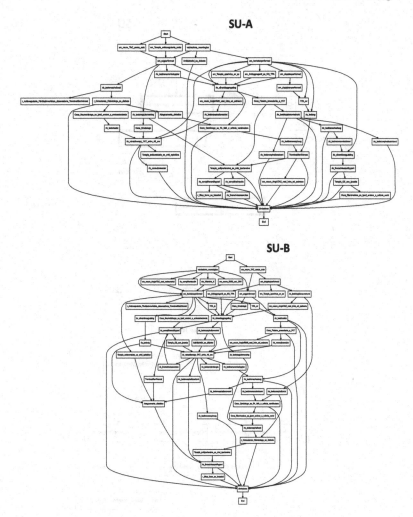

**Fig. 3.** Comparison between two process models, mined by resorting to Heuristic Miner, operating on ground traces. The figure is not intended to be readable, but only to give an idea of how complex the models can be

However, the model for SU-A at the top of Fig. 4 exhibits a more complex control flow (with the presence of loops), and shows three additional macro-actions with respect to the model of SU-B, namely "Extracranial Vessel Inspection", "Intracranial Vessel Inspection" and "Recanalization". This finding can be explained, since SU-A is a better equipped SU, where different kinds of patients, including some atypical/more critical ones, can be managed, thanks to the availability of different skills and instrumental resources. These patients may require the additional macro-actions reported in the model, and/or the repetition of some procedures, in order to better characterize and manage the individual patient's situation.

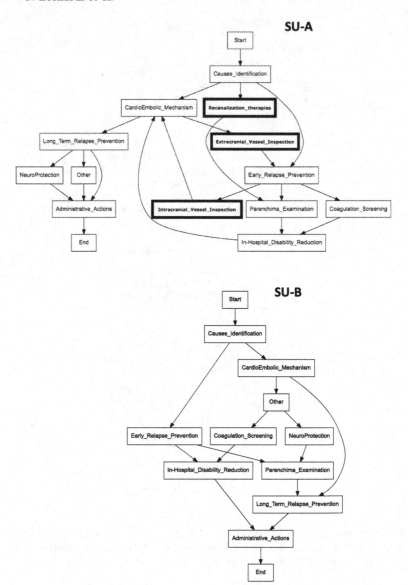

**Fig. 4.** Comparison between the two process models of the same SUs as Fig. 3, mined on abstracted traces. Additional macro-actions executed at SU-A are highlighted in bold

On the other hand, SU-B is a more generalist SU, where very specific human knowledge or technical resources are missing. As a consequence, the overall model control flow is simpler, and some activities are not executed at all.

Interestingly, our abstraction mechanism, while hiding irrelevant details, allows to still appreciate these differences.

## 4.2   Validation

For the validation study, we asked a SUN stroke management expert to provide a ranking of some SUN stroke units (see Table 1, column 2), on the basis of the quality of service they provide, with respect to the top level SU (referred as H0 in the experiments). Such a ranking was based on her personal knowledge of the SUs human and instrumental resource availability (not on the process models); therefore, it was qualitative, and *coarse-grained*, in the sense that more than one SU could obtain the same qualitative evaluation. The expert identified 6 SUs (H1–H6) with a high similarity level with respect to H0; 5 SUs (H7–H11) with a medium similarity level with respect to H0; and 4 SUs (H12–H15) with a low similarity level with respect to H0. The ordering of the SUs within one specific similarity level is not very relevant, since, as observed, the expert's ranking is coarse-grained. It is instead important to distinguish between different similarity levels.

We mined the process models of the 16 SUs by resorting to Heuristic Miner, both working on ground traces, and working on abstracted traces. We then ordered the two available process model sets with respect to H0, resorting to two different process model distances, i.e., the one described in [11], and the one presented in [22], globally obtaining four rankings.

These two distance measures are quite dissimilar. In particular, the distance in [11] provides a normalized version of the graph edit distance [5] for comparing business process models, and defines syntactical edit operation costs for node substitution (relying on edit distance between action names), node insertion/deletion, and edge insertion/deletion. On the other hand, the distance in [22] moves towards a more semantic and knowledge-intensive approach: when actions are represented in an ontology, as it is our case, it can adopt Palmer's distance [23] for calculating action node substitution costs. Palmer's distance between two actions is set to the normalized number of arcs on the path between the two actions themselves in the ontology. So, given the semantics of our ontology, two different actions can be considered as more or less distant on the basis of their goal. Moreover, the distance in [22] also considers edge substitution, which is disregarded in [11]. Indeed, Heuristic Miner labels edges with information that can be relevant in graph comparison, such as reliability [35], while statistical temporal information can be easily extracted from the log and saved as arc properties.

We exploited both these metrics because we wished to verify the effect of trace abstraction independently of the simplicity or completeness of the distance definition.

Results are shown in Table 1.

Column 1 in Table 1 shows the levels of similarity with respect to the reference SU. Column 2 shows the ranking according to the human medical expert; columns 3 and 4 show the ranking obtained by relying on the distance in [11], mining the process models on ground and abstracted traces, respectively. Similarly, columns 5 and 6 show the results obtained by relying on the distance in [22], mining the process models on ground and abstracted traces, respectively. In particular, abstraction was conducted at level 1 in the ontology (where level 0 is the root).

When working on ground traces, the distance in [11] correctly rates two process models in the high similarity group (33%), one process model in the medium similarity group (20%), and one process model in the low similarity group (25%, column 3). When working on abstracted traces, on the other hand, the distance in [11] correctly rates three process models in the high similarity group (50%), one process models in the medium similarity group (20%), and one process model in the low similarity group (25%, column 4).

When working on ground traces, the distance in [22] correctly rates two process models in the high similarity group (33%), zero process model in the medium similarity group (0%), and one process model in the low similarity group (25%, column 5). When working on abstracted traces, on the other hand, the distance in [22] correctly rates four process models in the high similarity group (66%), two process models in the medium similarity group (40%), and two process model in the low similarity group (50%, column 6).

In summary, when working on abstracted traces, both distances lead to rankings that are closer to the qualitative ranking provided by the human expert, but the improvement is larger when adopting the distance in [22], probably because it is able to take into account semantic information. We plan to verify this statement by means of further experiments in the future.

For the sake of completeness, we have also tested the effect of abstraction by evaluating the capability of the two distances in locating the high level SUs

**Table 1.** Ordering of 15 SUs, with respect to a given query model. Correct positions in the rankings with respect to the expert's qualitative similarity levels are highlighted in bold.

| Similarity | Medical expert | [11] ground | [11] abs | [22] ground | [22] abs |
|---|---|---|---|---|---|
| High | H1 | H15 | **H6** | **H3** | **H1** |
| High | H2 | H8 | H13 | **H2** | **H6** |
| High | H3 | **H5** | **H2** | H15 | **H5** |
| High | H4 | **H6** | **H4** | H10 | **H3** |
| High | H5 | H12 | H10 | H8 | H8 |
| High | H6 | H11 | H8 | H11 | H10 |
| Medium | H7 | H4 | H3 | H4 | **H9** |
| Medium | H8 | H3 | H5 | H12 | H4 |
| Medium | H9 | H14 | **H7** | H6 | H14 |
| Medium | H10 | H2 | H14 | H1 | **H7** |
| Medium | H11 | **H10** | H15 | H13 | H12 |
| Low | H12 | **H13** | H9 | H9 | **H13** |
| Low | H13 | H9 | H1 | **H14** | H2 |
| Low | H14 | H1 | H11 | H7 | H11 |
| Low | H15 | H7 | **H12** | H5 | **H15** |

(i.e., the SUs with a high similarity with respect to H0) as the first six items of the overall ranking. To this end, we calculated the $nDCG_6$[7] index on the rankings of Table 1. Also in this experiment, we were able to verify that the use of abstraction leads to results that are closer to the (ideal) ones provided by the human expert (see Table 2), with a larger improvement when adopting the distance in [22].

**Table 2.** $nDCG_6$ index calculation on the rankings provided by the two metrics.

| [11] ground | [11] abs | [22] ground | [22] abs |
|---|---|---|---|
| 0.621 | 0.798 | 0.781 | 0.925 |

Table 3, on the other hand, reports our results on the calculation of *fitness* [29] on the process models mined for our 16 SUs, at different levels of abstraction. Fitness evaluates whether a process model is able to reproduce all execution sequences that are in the log. If the log can be replayed correctly, fitness evaluates to 1. In the worst case, it evaluates to 0. Fitness calculation is available in ProM.

**Table 3.** Fitness values calculated on the mined process models, when operating at different levels of abstraction

| SU | Ground | Abs. level 2 | Abs. level 1 |
|---|---|---|---|
| H0 | 0.58 | 0.83 | 0.87 |
| H1 | 0.43 | 0.73 | 0.97 |
| H2 | 0.47 | 0.72 | 0.87 |
| H3 | 0.40 | 0.85 | 0.95 |
| H4 | 0.42 | 0.91 | 0.93 |
| H5 | 0.42 | 0.83 | 0.87 |
| H6 | 0.52 | 0.76 | 0.95 |
| H7 | 0.52 | 0.79 | 0.92 |
| H8 | 0.75 | 0.78 | 0.92 |
| H9 | 0.46 | 0.85 | 0.92 |
| H10 | 0.41 | 0.83 | 0.87 |
| H11 | 0.44 | 0.61 | 0.90 |
| H12 | 0.49 | 0.67 | 0.89 |
| H13 | 0.46 | 0.84 | 0.91 |
| H14 | 0.43 | 0.85 | 0.90 |
| H15 | 0.54 | 0.78 | 0.90 |

As it can be observed from the table, the more the traces are abstracted, the more the fitness values increase in the corresponding mined models.

In conclusion, our abstraction mechanism, while hiding irrelevant details, allows to still appreciate relevant differences between models, such as, e.g., the presence/absence of important actions, as commented in Sect. 4.1. Moreover, when working on abstracted traces, the adoption of different distance definitions leads to SU rankings that are closer to the qualitative ranking provided by the human expert. Finally, very interestingly, abstraction proves to be a means to significantly increase the quality of the mined models, measured in terms of fitness, which is a well known and largely adopted indicator.

## 5 Related Works

The use of semantics in business process management, with the aim of operating at different levels of abstractions in process discovery and/or analysis, is a relatively young area of research, where much is still unexplored.

One of the first contributions in this field was proposed in [6], which introduces a process data warehouse, where taxonomies are exploited to add semantics to process execution data, in order to provide more intelligent reports. The work in [14] extends the one in [6], presenting a complete architecture that allows business analysts to perform multidimensional analysis and classify process instances, according to flat taxonomies (i.e., taxonomies without subsumption relations between concepts).

Hepp et al. [17] propose a framework able to merge semantic web, semantic web services, and business process management techniques to build semantic business process management, and use ontologies to provide machine-processable semantics in business processes [18]. The work in [26] develops in a similar context, and extends OLAP tools with semantics (exploiting ontologies rather than (flat) taxonomies).

The topic was studied in the SUPER project [25], within which several ontologies were created, such as the process mining ontology and the event ontology [24]; these ontologies define core terminologies of business process management, usable by machines for task automation. However, the authors did not present any concrete implementations of semantic process mining or analysis.

Ontologies, references from elements in logs to concepts in ontologies, and ontology reasoners (able to derive, e.g., concept equivalence), are described as the three essential building blocks for semantic process mining in [10]. This paper also shows how to use these building blocks to extend ProM's LTL Checker [31] to perform semantic auditing of logs.

The work in [8] focuses on the use of semantics in business process monitoring, an activity that allows to detect or predict process deviations and special situations, to diagnose their causes, and possibly to resolve problems by applying corrective actions. Detection, diagnosis and resolution present interesting challenges that, on the authors' opinion, can strongly benefit from knowledge-based techniques.

In [8,9] the idea to explicitly relate (or annotate) elements in the log with the concepts they represent, linking these elements to concepts in ontologies, is addressed. This "semantic lifting" approach, to use a term borrowed from the Web scenario, is also investigated in [3], to discover the process that is actually being executed.

In [9] an example of process discovery at different levels of abstractions is presented. It is however a very simple example, where a couple of ground actions are abstracted according to their common ancestor. However, the management of interleaved actions or delays is not addressed, and multiple inheritance is not considered. A more recent work [19] introduces a methodology that combines domain and company-specific ontologies and databases to obtain multiple levels of abstraction for process mining. In this paper data in databases become instances of concepts at the bottom level of a taxonomy tree structure. If consecutive tasks in the discovered model abstract as the same concepts, those tasks are aggregated. However, also in this work we could find neither a clear description of the abstraction algorithm, nor the management of interleaved actions or delays.

Moreover, most of the papers cited above (including [9,10]) present theoretical frameworks, and not yet a detailed technical architecture nor a concrete implementation of all their ideas.

Referring to medical applications, the work in [13] proposes an approach, based on semantic process mining, to verify the compliance of a Computer Interpretable Guideline with medical recommendations. In this case, semantic process mining refers to conformance checking rather than to process discovery (as it is also the case in [10]). These works are thus only loosely related to our contribution.

Besides the contributions listed above, most of which can be categorized as normative/deductive approaches, it is worth citing some interesting emergent approaches as well [4,27]. The work in [27] affords the problem of dealing with very big log files, proposing solutions for scalable process discovery and extending XES [34], a log file format introduced to solve problems with the semantics of additional attributes; the work in [4] moves even forward, presenting a methodology designed to implement consistent process mining algorithms in a Big Data context, managing semantic heterogeneity.

Another interesting research direction adopts abstraction as a way to establish a relationship between the events typically recorded in the log by the information system, and the high-level actions which are of interest when mining business process models. In particular the work in [12] looks for mappings between events and actions. The set of possible mappings can be large, and one should focus on the mapping which has the highest coverage, meaning that the chosen mapping should be applicable to the largest possible number of traces in the log. The range of a mapping, i.e., the set of actions that appear in the mapping, is also a very important factor. In particular, the authors adopt a greedy approach, where they try to merge those mappings which, by themselves, already have the largest range and highest coverage among their peers. The work in [28], on the other hand, shows that supervised learning (namely Condition Random Fields) can be leveraged for the event abstraction task when annotations with high-level interpretations of the low-level events are available for a subset of the traces. Conditional Random Fields are trained on the annotated traces to create a probabilistic mapping from low-level events to high-level actions. This mapping, once obtained, can be applied to the unannotated traces as well. In [21], the authors align the behavior defined by activity patterns with the observed behavior in the log. Activity patterns encode assumptions on how high-level actions manifest themselves in terms of recorded low-level events. The work in [16] exploits trace segmentation. Trace segmentation is based on the idea that subsequences of events, which are supposed to be the product of a higher-level action, are identified. From the co-occurrence of events in the log, the authors derive the relative correlation between their event classes. All event classes found in the log are successively combined into clusters, representing higher-level types. In this hierarchy of event classes, an arbitrary level of abstraction can be chosen to process the log.

Most of these constributions (see, e.g., [12,16]), however, adopt non-semantic approaches. Moreover, as stated in the Introduction, in our work we assume that traces are sequences of actions (i.e., with respect to event traces, they have already been pre-processed). Therefore, this line of research is only loosely related to ours.

In conclusion, in the current research panorama, our work appears to be very innovative, for several reasons:

- many approaches, illustrating very interesting and sometimes ambitious ideas, just provide pure theoretical frameworks, which can be very important to inspire more engineering-style work. However, concrete implementations of algorithms and complete architectures of systems are often missing, leaving open research opportunities for contributions like the one we have presented;
- in semantic process mining, more work has been done in the field of conformance checking (also in medical applications), while process discovery still deserves attention (also because many approaches are still at the theoretical level, as commented above);
- as regards trace abstraction, it is often proposed as a very powerful means to obtain better process discovery and analysis results, but technical details of the abstraction mechanism are usually not provided, or are illustrated through very simple examples, where the issues related to the management of interleaved actions or delays do not emerge.

## 6    Concluding Remarks and Future Work

In this paper, we have presented a framework for knowledge-based abstraction of log traces. In our approach, abstracted traces are then provided as an input to semantic process mining. Semantic process mining relies on ProM algorithms; indeed, the overall integration of our approach within ProM is foreseen in our future work.

Our case study in the field of stroke management suggests that the capability of abstracting the log traces on the basis of their semantic goal allows to mine clearer process models, where unnecessary details are hidden, but key behaviors are clear.

Moreover, in our validation study we mined the process models of some SUs by resorting to Heuristic Miner, both working on ground traces, and working on abstracted traces. We then ordered the two available process model sets with respect to the model of the best equipped SU in the SUN network, resorting to two different process model distances, i.e., the one described in [11], and the one presented in [22], globally obtaining four rankings. We verified that, when working on abstracted traces, both distances lead to rankings that are closer to the qualitative ranking provided by a domain expert, but the improvement is much larger when adopting the distance in [22], probably because it is able to take into account semantic information. Finally, abstraction proves to be a means to significantly increase the quality of the mined models, when measured in terms of fitness.

In the future, we would like to test the approach in different application domains as well, even if we know that the task will require time and may present some issues. Indeed, domain knowledge acquisition is typically time consuming: our work on formalizing the stroke ontology and the rules required multiple sessions of work with physicians. Moreover, the domain of stroke management can count on internationally recognized guidelines, which are well detailed: therefore, the acquired knowledge can be considered as comprehensive and accurate; this may not hold in different domains. The domain choice will therefore be critical, and possibly we will select a non medical application.

We also plan to apply the abstraction mechanism to subsets of the log, obtained by clustering techniques or possibly by domain expert rating, in order to test the effect of abstraction on a more homogeneous input.

The quality of the mined process models could also be evaluated by resorting to conformance checking techniques.

We will also consider how to cobine our approach with the pattern-based abstraction described in [21].

Finally, an abstraction mechanism directly operating on process models (i.e., on the graph, instead of the log), may be considered, possibly along the lines described in [15], and abstraction results will be compared to the ones currently enabled by our framework.

# References

1. Aguilar, E.R., Ruiz, F., García, F., Piattini, M.: Evaluation measures for business process models. In: Haddad, H. (ed.) Proceedings of the 2006 ACM Symposium on Applied Computing (SAC), Dijon, France, 23–27 April 2006, pp. 1567–1568. ACM (2006)
2. Allen, J.F.: Towards a general theory of action and time. Artif. Intell. **23**, 123–154 (1984)

3. Azzini, A., Braghin, C., Damiani, E., Zavatarelli, F.: Using semantic lifting for improving process mining: a data loss prevention system case study. In: Accorsi, R., Ceravolo, P., Cudré-Mauroux, P. (eds.) Proceedings of the 3rd International Symposium on Data-Driven Process Discovery and Analysis, CEUR Workshop Proceedings, vol. 1027, pp. 62–73. CEUR-WS.org (2013)

4. Azzini, A., Ceravolo, P.: Consistent process mining over big data triple stores. In: IEEE International Congress on Big Data, BigData Congress 2013, pp. 54–61. IEEE Computer Society (2013)

5. Bunke, H.: On a relation between graph edit distance and maximum common subgraph. Pattern Recogn. Lett. **18**(8), 689–694 (1997)

6. Casati, F., Shan, M.-C.: Semantic analysis of business process executions. In: Jensen, C.S., et al. (eds.) EDBT 2002. LNCS, vol. 2287, pp. 287–296. Springer, Heidelberg (2002). https://doi.org/10.1007/3-540-45876-X_19

7. Croft, W.B., Metzler, D., Strohman, T.: Search Engines: Information Retrieval in Practice. Alternative Etext Formats. Addison-Wesley, Boston (2010)

8. de Medeiros, A.K.A., et al.: An outlook on semantic business process mining and monitoring. In: Meersman, R., Tari, Z., Herrero, P. (eds.) OTM 2007. LNCS, vol. 4806, pp. 1244–1255. Springer, Heidelberg (2007). https://doi.org/10.1007/978-3-540-76890-6_52

9. Alves de Medeiros, A.K., van der Aalst, W.M.P.: Process mining towards semantics. In: Dillon, T.S., Chang, E., Meersman, R., Sycara, K. (eds.) Advances in Web Semantics I. LNCS, vol. 4891, pp. 35–80. Springer, Heidelberg (2008). https://doi.org/10.1007/978-3-540-89784-2_3

10. Alves de Medeiros, A.K., van der Aalst, W.M.P., Pedrinaci, C.: Semantic process mining tools: core building blocks. In: Golden, W., Acton, T., Conboy, K., van der Heijden, H., Tuunainen, V.K. (eds.) 16th European Conference on Information Systems, ECIS 2008, Galway, Ireland, pp. 1953–1964 (2008)

11. Dijkman, R., Dumas, M., García-Bañuelos, L.: Graph matching algorithms for business process model similarity search. In: Dayal, U., Eder, J., Koehler, J., Reijers, H.A. (eds.) BPM 2009. LNCS, vol. 5701, pp. 48–63. Springer, Heidelberg (2009). https://doi.org/10.1007/978-3-642-03848-8_5

12. Ferreira, D.R., Szimanski, F., Ralha, C.G.: Improving process models by mining mappings of low-level events to high-level activities. J. Intell. Inf. Syst. **43**(2), 379–407 (2014)

13. Grando, M.A., Schonenberg, M.H., van der Aalst, W.M.P.: Semantic process mining for the verification of medical recommendations. In: Traver, V., Fred, A.L.N., Filipe, J., Gamboa, H. (eds.) HEALTHINF 2011 - Proceedings of the International Conference on Health Informatics, Rome, Italy, 26–29 January 2011, pp. 5–16. SciTePress (2011)

14. Grigori, D., Casati, F., Castellanos, M., Dayal, U., Sayal, M., Shan, M.C.: Business process intelligence. Comput. Ind. **53**(3), 321–343 (2004)

15. Günther, C.W., van der Aalst, W.M.P.: Fuzzy mining – adaptive process simplification based on multi-perspective metrics. In: Alonso, G., Dadam, P., Rosemann, M. (eds.) BPM 2007. LNCS, vol. 4714, pp. 328–343. Springer, Heidelberg (2007). https://doi.org/10.1007/978-3-540-75183-0_24

16. Günther, C.W., Rozinat, A., van der Aalst, W.M.P.: Activity mining by global trace segmentation. In: Rinderle-Ma, S., Sadiq, S., Leymann, F. (eds.) BPM 2009. LNBIP, vol. 43, pp. 128–139. Springer, Heidelberg (2010). https://doi.org/10.1007/978-3-642-12186-9_13

17. Hepp, M., Leymann, F., Domingue, J., Wahler, A., Fensel, D.: Semantic business process management: a vision towards using semantic web services for business process management. In: Lau, F.C.M., Lei, H., Meng, X., Wang, M. (eds.) 2005 IEEE International Conference on e-Business Engineering (ICEBE 2005), 18–21 October 2005, Beijing, China, pp. 535–540. IEEE Computer Society (2005)

18. Hepp, M., Roman, D.: An ontology framework for semantic business process management. In: Oberweis, A., Weinhardt, C., Gimpel, H., Koschmider, A., Pankratius, V., Schnizler, B. (eds.) eOrganisation: Service-, Prozess-, Market-Engineering: 8. Internationale Tagung Wirtschaftsinformatik - Band 1, WI 2007, Karlsruhe, Germany, 28 February–2 March 2007, pp. 423–440. Universitaetsverlag Karlsruhe (2007)

19. Jareevongpiboon, W., Janecek, P.: Ontological approach to enhance results of business process mining and analysis. Bus. Process Manag. J. **19**(3), 459–476 (2013)

20. De Maio, M.N., Salatino, M., Aliverti, E.: Mastering JBoss Drools 6 for Developers. Packt Publishing, Birmingham (2016)

21. Mannhardt, F., de Leoni, M., Reijers, H.A., van der Aalst, W.M.P., Toussaint, P.J.: From low-level events to activities - a pattern-based approach. In: La Rosa, M., Loos, P., Pastor, O. (eds.) BPM 2016. LNCS, vol. 9850, pp. 125–141. Springer, Cham (2016). https://doi.org/10.1007/978-3-319-45348-4_8

22. Montani, S., Leonardi, G., Quaglini, S., Cavallini, A., Micieli, G.: Improving structural medical process comparison by exploiting domain knowledge and mined information. Artif. Intell. Med. **62**(1), 33–45 (2014)

23. Palmer, M., Wu, Z.: Verb semantics for English-Chinese translation. Mach. Transl. **10**, 59–92 (1995)

24. Pedrinaci, C., Domingue, J.: Towards an ontology for process monitoring and mining. In: Hepp, M., Hinkelmann, K., Karagiannis, D., Klein, R., Stojanovic, N. (eds.) Proceedings of the Workshop on Semantic Business Process and Product Lifecycle Management SBPM 2007, held in conjunction with the 3rd European Semantic Web Conference (ESWC 2007), Innsbruck, Austria, 7 June 2007, vol. 251. CEUR Workshop Proceedings (2007)

25. Pedrinaci, C., Domingue, J., Brelage, C., van Lessen, T., Karastoyanova, D., Leymann, F.: Semantic business process management: scaling up the management of business processes. In: Proceedings of the 2th IEEE International Conference on Semantic Computing (ICSC 2008), 4–7 August 2008, Santa Clara, California, USA, pp. 546–553. IEEE Computer Society (2008)

26. Sell, D., Cabral, L., Motta, E., Domingue, J., dos Santos Pacheco, R.C.: Adding semantics to business intelligence. In: 16th International Workshop on Database and Expert Systems Applications (DEXA 2005), 22–26 August 2005, Copenhagen, Denmark, pp. 543–547. IEEE Computer Society (2005)

27. Syamsiyah, A., van Dongen, B.F., van der Aalst, W.M.P.: DB-XES: enabling process discovery in the large. In: Ceravolo, P., Guetl, C., Rinderle-Ma, S. (eds.) SIMPDA 2016. LNBIP, vol. 307, pp. 53–77. Springer, Cham (2018). https://doi.org/10.1007/978-3-319-74161-1_4

28. Tax, N., Sidorova, N., Haakma, R., van der Aalst, W.: Event abstraction for process mining using supervised learning techniques. CoRR, abs/1606.07283 (2016)

29. van der Aalst, W.: Process Mining. Data Science in Action. Springer, Heidelberg (2016). https://doi.org/10.1007/978-3-662-49851-4

30. van der Aalst, W., van Dongen, B., Herbst, J., Maruster, L., Schimm, G., Weijters, A.: Workflow mining: a survey of issues and approaches. Data Knowl. Eng. **47**, 237–267 (2003)

31. van der Aalst, W.M.P., de Beer, H.T., van Dongen, B.F.: Process mining and verification of properties: an approach based on temporal logic. In: Meersman, R., Tari, Z. (eds.) OTM 2005. LNCS, vol. 3760, pp. 130–147. Springer, Heidelberg (2005). https://doi.org/10.1007/11575771_11

32. van Dongen, B., Alves De Medeiros, A., Verbeek, H., Weijters, A., van der Aalst, W.: The ProM framework: a new era in process mining tool support. In: Ciardo, G., Darondeau, P. (eds.) Knowl. Mang. Integr. Elem., pp. 444–454. Springer, Berlin (2005)

33. Vanderfeesten, I., Reijers, H.A., Mendling, J., van der Aalst, W.M.P., Cardoso, J.: On a quest for good process models: the cross-connectivity metric. In: Bellahsène, Z., Léonard, M. (eds.) CAiSE 2008. LNCS, vol. 5074, pp. 480–494. Springer, Heidelberg (2008). https://doi.org/10.1007/978-3-540-69534-9_36

34. Verbeek, H.M.W., Buijs, J.C.A.M., van Dongen, B.F., van der Aalst, W.M.P.: XES, XESame, and ProM 6. In: Soffer, P., Proper, E. (eds.) CAiSE Forum 2010. LNBIP, vol. 72, pp. 60–75. Springer, Heidelberg (2011). https://doi.org/10.1007/978-3-642-17722-4_5

35. Weijters, A., van der Aalst, W., Alves de Medeiros, A.: Process Mining with the Heuristic Miner Algorithm, WP 166. Eindhoven University of Technology, Eindhoven (2006)

# Mining Local Process Models
# and Their Correlations

Laura Genga[(✉)] [ID], Niek Tax, and Nicola Zannone [ID]

Eindhoven University of Technology, 5600 Eindhoven, MB, The Netherlands
{l.genga,n.tax,n.zannone}@tue.nl

**Abstract.** Mining local patterns of process behavior is a vital tool for the analysis of event data that originates from flexible processes, which in general cannot be described by a single process model without overgeneralizing the allowed behavior. Several techniques for mining local patterns have been developed over the years, including Local Process Model (LPM) mining, episode mining, and the mining of frequent subtraces. These pattern mining techniques can be considered to be orthogonal, i.e., they provide different types of insights on the behavior observed in an event log. In this work, we demonstrate that the joint application of LPM mining and other patter mining techniques provides benefits over applying only one of them. First, we show how the output of a subtrace mining approach can be used to mine LPMs more efficiently. Secondly, we show how instances of LPMs can be correlated together to obtain larger LPMs, thus providing a more comprehensive overview of the overall process. We demonstrate both effects on a collection of real-life event logs.

## 1 Introduction

*Process Mining* [1] has emerged as a new field that aims at business process improvement through the analysis of *event logs* recorded by information systems. Such event logs capture the different steps (events) that are recorded for each instance of the process, and record for each event what was done, by whom, for whom, where, when, etc. One of the main challenges within process mining is *process discovery*, where the aim is to discover an interpretable and accurate model of the process based on an event log. The resulting process model provides insight into what is happening in the process and can be used as a starting point for more in-depth process analysis, e.g., bottleneck analysis [26], and checking compliance with rules and regulations [27].

In recent years, the scope of process discovery has broadened to novel application domains, such as software analysis and human behavior analysis. In some of those new application domains the logs have a *high degree of variability*, thereby making it difficult to represent the behavior observed in the log in a process model. High log variability significantly impacts the generation of insightful models; the process models obtained using process discovery techniques often do not provide useful insights into the process behavior, either because they overgeneralize, thus tending to allow for any sequence of events (e.g., [20,35]),

© IFIP International Federation for Information Processing 2019
Published by Springer Nature Switzerland AG 2019
P. Ceravolo et al. (Eds.): SIMPDA 2017, LNBIP 340, pp. 65–88, 2019.
https://doi.org/10.1007/978-3-030-11638-5_4

or because, on the contrary, they represent exactly all (or most of) the behaviors recorded in the log, thus providing a spaghetti-like representation that is typically too complex to be exploited by a human analyst (e.g., [7]).

Several techniques aim to address this challenge of analyzing highly variable event logs. *Declarative process discovery* (e.g., [23,29]) focuses on the mining of binary relations between activities of the process. *Local Process Model (LPM) mining* (e.g., [32,33]) aims at the mining of a collection of process models instead of a single model, where each model captures a subset of the process behavior. *Subtrace mining* (e.g., [3,9,19]) mines subtraces that represent relevant sequential portions of process executions (i.e., subprocesses). In this work we will focus on subtrace and LPM mining. These techniques share similar goals, i.e., the mining of relevant process execution patterns. However, they provide different insights on the process and have their advantages and disadvantages.

Subtrace mining techniques derive frequent patterns of sequential executions of process activities from event logs. Diamantini et al. [9] extend subtrace mining to discover partial order relations between process activities by either relying on a priori knowledge on concurrency relations or on concurrency detection mechanisms provided by process discovery techniques. However, subtrace mining techniques are not able to capture control-flow constructs other than sequential and concurrency relations between process activities. Rather, some approaches focus on relations *between patterns* instead of between activities from the process. For instance, the work in [9] constructs hierarchies of patterns where subtraces are ordered with respect to the inclusion relation. Genga et al. [13] apply frequent itemset mining techniques to mine partial order relations between subtraces.

Local Process Model (LPM) mining aims at mining process patterns that can describe any arbitrary combination of sequential ordering, concurrency, loops and choice construct. However, mining LPM patterns is computationally expensive, or even infeasible, for event logs with many activities. In practice, computational problems can already arise at seventeen activities [32]. Therefore, a set of heuristics have been proposed in [32] to speed up the mining process. These heuristics discover subsets of process activities (called *projections*) that are strongly related and apply the LPM miner to each projection individually, aggregating the results by taking the union of the resulting LPMs. The downside of these heuristics is the loss of formal guarantees that all frequent local process models are found.

In this work, we explore the synergies between subtrace mining and LPM mining in two ways. First, we investigate the application of the patterns obtained using subtrace mining for LPM mining. This subtrace-based LPM mining approach generates *projections* based on the subtraces mined using the technique presented in [9] and furthermore extracts *ordering constraints* from the subtraces to reduce the search space of LPM mining. We conjecture that using activities from these subtraces as projections and ordering constraints can speed up the LPM mining procedure. Secondly, we explore the application of approaches to mine higher level relations *between* subtraces to generate larger LPMs. In particular, these approaches allow us to merge LPMs describing possibly unconnected portions of the process behavior, providing a more comprehensive overview of the overall process, which would otherwise be difficult to achieve using original LPM algorithms due to the large number of activities involved.

This paper is organized as follows. Section 2 introduces notation and basic concepts that are used throughout the paper. Section 3 presents a projection method and a constraint generation technique for LPM mining, while Sect. 4 presents a method to infer ordering relations between mined LPMs. In Sect. 5 we evaluate both techniques on a collection of real-life event logs. Finally, Sect. 6 discusses related work and Sect. 7 concludes the paper.

## 2  Background

In this section we introduce notation and basic concepts used throughout the paper. We start by introducing event data and process models in Sect. 2.1 and then we introduce methods for mining subprocess models from event logs in Sect. 2.2.

### 2.1  Event Data and Process Models

Process models describe how processes should be carried out. Two process model notations that are commonly used in process mining are *process trees* [5] and *Petri nets* [28]. A process tree is a tree structure where leaf nodes represent process activities, while non-leaf nodes represent *operators* that specify the allowed behavior over the activity nodes. Allowed operator nodes are the *sequence* operator ($\rightarrow$), which indicates that the first child is executed before the second, the *exclusive choice* operator ($\times$), which indicates that exactly one of the children can be executed, the *concurrency* operator ($\wedge$), which indicates that every child will be executed but allows for any ordering, and the *loop* operator ($\circlearrowleft$), which has one child node and allows for repeated execution of this node.

We formally define process trees recursively. Let $\Sigma$ be the set of all process activities, $OP = \{\rightarrow, \times, \wedge, \circlearrowleft\}$ a set of operators and symbol $\tau \notin \Sigma$ denotes silent activities. We define a process tree $pt$ as follows:

- $a \in \Sigma \cup \{\tau\}$ is a process tree $M$;
- let $\{M_1, M_2, \ldots, M_n\}$ be a set of process trees. Then $\oplus(M_1, M_2, \ldots, M_n)$ with $\oplus \in OP$ is a process tree.

Hereafter, $\mathfrak{L}(M)$ denotes the language of a process model $M$, i.e., the set of activity execution paths allowed by the model. Figure 1d shows an example process tree $M_4$, with $\mathfrak{L}(M_4)=\{\langle a, b, c\rangle, \langle a, c, b\rangle, \langle d, b, c\rangle, \langle d, c, b\rangle\}$. Informally, it indicates that either activity $a$ or $d$ is executed first, followed by the execution of activities $b$ and $c$ in any order.

A *Petri net* $N = \langle P, T, F, \ell \rangle$ is a tuple where $P$ is a finite set of *places*, $T$ is a finite set of *transitions* such that $P \cap T = \emptyset$, $F \subseteq (P \times T) \cup (T \times P)$ is a set of directed arcs, called the flow relation, and $\ell : T \nrightarrow \Sigma$ is a labeling function that

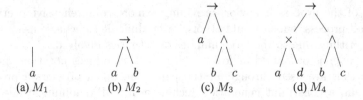

**Fig. 1.** An initial LPM ($M_1$) and three LPMs built from successive expansions.

assigns process activities to transitions. Unlabeled transitions, i.e., $t \in T$ with $t \notin dom(l)$, are referred to as $\tau$-transitions, or invisible transitions.

The state of a Petri net is defined by its *marking*. The marking assigns a finite number of tokens to each place. Transitions of the Petri net represent activities. The input places of a transition $t \in T$ are all places for which there is a directed edge to the transition, i.e. $\{p \in P | (p, t) \in F\}$. The output places of a transition are defined similarly as $\{p \in P | (t, p) \in F\}$. Executing a transition consumes one token from each of its input places and produces one token on each of its output places. A transition can only be executed when there is at least one token in each of its input places. Often we consider a Petri net in combination with an initial marking and a final marking, allowing us to define language $\mathcal{L}(N)$, consisting of all possible sequences of visible transition labels (i.e., ignoring $\tau$-transitions) that start in the initial marking and end in the final marking. It is worth noting that process trees can trivially be transformed into Petri nets.

Process discovery aims to mine a process model from past process executions. An *event* $e$ is the actual recording of the occurrence of an activity in $\Sigma$. A *trace* $\sigma$ is a sequence of events, i.e., $\sigma = \langle e_1, e_2, \ldots, e_n \rangle \in \Sigma^*$. An *event log* $L \in \mathbb{N}^{\Sigma^*}$ is a finite multiset of traces. For example, event log $L = [\langle a, b, c \rangle^2, \langle b, a, c \rangle^3]$ consists of two occurrences of trace $\langle a, b, c \rangle$ and three occurrences of trace $\langle b, a, c \rangle$. $L\!\restriction_X$ represents the projection of log $L$ on a subset of the activities $X \subseteq \Sigma$, e.g., $L\!\restriction_{\{b,c\}} = [\langle b, c \rangle^5]$. $\#(\sigma, L)$ denotes the frequency of sequence $\sigma \in \Sigma^*$ as a sub-trace within log $L$, e.g., $\#(\langle a, b \rangle, [\langle a, b, c \rangle^2, \langle a, b, d \rangle^3]) = 5$. $\sigma_1 \cdot \sigma_2$ denotes the concatenation of sequences $\sigma_1$ and $\sigma_2$, e.g., $\langle a, b \rangle \cdot \langle c, d, e \rangle = \langle a, b, c, d, e \rangle$.

## 2.2  Subprocess Mining

Two methods to mine subprocesses from event logs are *Local Process Models* (LPMs) [33] and *subtrace mining* [3,9,19]. LPMs are process models that describe frequent but partial behaviors seen in the event log, i.e., they model subsets of the process.

**LPM Mining.** [33] is a technique to generate a ranked collection of LPMs through iterative expansion of candidate process trees. This technique encompasses four steps: (1) the *generation* of an initial set of process trees, consisting of one process tree for each activity; (2) the *evaluation* phase, where process tree

| event id | activity | time |
|----------|----------|------|
| 1 | a | 15-4-2016 12:23 |
| 2 | d | 16-4-2016 14:38 |
| 3 | b | 16-4-2016 14:46 |
| 4 | c | 16-4-2016 15:46 |
| 5 | d | 16-4-2016 16:53 |
| 6 | c | 16-4-2016 16:58 |
| 7 | a | 16-4-2016 17:11 |
| 8 | c | 16-4-2016 17:45 |
| 9 | b | 16-4-2016 18:03 |
| 10 | d | 17-4-2016 12:09 |
| 11 | a | 17-4-2016 18:24 |
| 12 | b | 17-4-2016 18:36 |
| 13 | a | 17-4-2016 18:37 |

$$\sigma = \langle a,d,b,c,d,c,a,c,b,d,a,b,a \rangle$$
$$\sigma\restriction_{\{a,b,c\}} = \langle \underline{a,b,c},c,\underline{a,c,b},a,b,a \rangle$$
$$\phantom{xxxxxx} \lambda_1 \; \gamma_1 \; \lambda_2 \; \gamma_2 \; \lambda_3$$

$$\Gamma_{\sigma,LPM} = \langle a,b,c,a,c,b \rangle$$

(a) A trace $\sigma$ of an event log $L$      (b) Segmentation of $\sigma$ on $M_3$

**Fig. 2.** Example of segmentation in LPM mining.

quality is assessed by a set of tailored metrics; (3) the *selection* phase, where process trees that do not meet certain criteria are removed; (4) the *expansion* phase, where candidates selected at the previous step are expanded by replacing an activity node $a$ by an operator node ($\rightarrow$, $\times$, $\wedge$ or $\circlearrowright$), whose children are the replaced activity $a$ and another activity $b \in \Sigma$ of the process. Steps 2 to 4 are repeated until no new candidate meets the criteria.

An LPM $M$ can be expanded in many ways, as any one of its activity nodes can be replaced, using any of the operator nodes in combination with any other activity from the set of activities in the log. $Exp(M)$ denotes the set of expansions of $M$ (described in more detail in [33]), and $exp\_max$ the maximum number of expansions allowed from an *initial LPM*, i.e., the LPMs generated in step 1.

Figure 1 provides an example of the expansion procedure, starting from the initial LPM $M_1$ of Fig. 1a. The LPM of Fig. 1a is first expanded into a larger LPM by replacing $a$ by operator node $\rightarrow$, with activity $a$ as its left child node and $b$ as its right child node, resulting in the LPM of Fig. 1b. Note that $M_1$ can also be expanded using any other operator or any other activity from $\Sigma$, and LPM discovery recursively explores all possible process trees that meet a support threshold by iterative expansion. In a second expansion step, activity node $b$ of the LPM of Fig. 1b is replaced by operator node $\wedge$, with activity $b$ as its left child and $c$ as its right child, resulting in the LPM of Fig. 1c. Finally, activity node $a$ of the LPM of Fig. 1c is replaced by operator node $\times$ with activity $a$ as its left child and activity $d$ as its right child, forming the LPM of Fig. 1d. In traditional LPM discovery the expansion procedure of an LPM stops when the behavior described by the LPM is not observed frequently enough in an event log $L$ (i.e., with regard to some *support threshold*). LPMs are mined in process trees representation, but often their Petri net representation is used to visualize them.

To evaluate a given LPM on a given event log $L$, its traces $\sigma \in L$ are first projected on the set of activities $X$ in the LPM, i.e. $\sigma' = \sigma\restriction_X$. The projected trace $\sigma'$ is then segmented into $\gamma$-segments, i.e., segments that fit the behavior of the LPM, and $\lambda$-segments, i.e. segments that do not fit the behavior of the

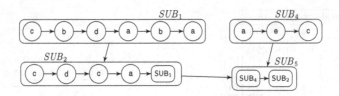

**Fig. 3.** Example of SUBDUE hierarchy

LPM. Specifically, $\sigma' = \lambda_1 \cdot \gamma_1 \cdot \lambda_2 \cdot \gamma_2 \ldots \lambda_n \cdot \gamma_n \cdot \lambda_{n+1}$ such that $\gamma_i \in \mathcal{L}(LPM)$ and $\lambda_i \notin \mathcal{L}(LPM)$. We define $\Gamma_{\sigma,LPM}$ to be a function that projects trace $\sigma$ on the LPM activities and obtains its subsequences that fit the LPM, i.e. $\Gamma_{\sigma,LPM} = \gamma_1 \cdot \gamma_2 \ldots \gamma_n$.

Let our LPM $M_3$ under evaluation be the process tree of Fig. 1c and $\sigma$ the example trace shown in Fig. 2a. Function $Act(LPM)$ gives the set of process activities in the LPM, e.g. $Act(M_3) = \{a, b, c\}$. The projection on the activities of the LPM gives $\sigma\lceil_{Act(M_3)} = \langle a, b, c, c, a, c, b, a, b, a\rangle$. Figure 2b shows the segmentation of the projected trace on the LPM, leading to $\Gamma_{\sigma,LPM} = \langle a, b, c, a, c, b\rangle$. The segmentation starts with an empty non-fitting segment $\lambda_1$, followed by a fitting segment $\gamma_1 = \langle a, b, c\rangle$, which completes one run through the process tree. The second event $c$ in $\sigma$ cannot be replayed on LPM, since it only allows for one $c$ and $\gamma_1$ already contains a $c$. This results in a non-fitting segment $\lambda_2 = \langle c\rangle$. Segment $\gamma_2 = \langle a, c, b\rangle$ again represents a run through the process tree; the segmentation ends with non-fitting segment $\lambda_3 = \langle a, b, a\rangle$. We lift segmentation function $\Gamma$ to event logs, $\Gamma_{L,LPM} = \{\Gamma_{\sigma,LPM} | \sigma \in L\}$. An alignment-based [2] implementation of $\Gamma$, as well as a method to rank and select LPMs based on their support, i.e., the number of events in $\Gamma_{L,LPM}$, is described in [33].

LPMs only contain a subset of the activities of a log $L$, and therefore, each LPM $M$ can in principle be discovered on any projection on $L$ containing the activities used in $M$. The computational complexity of LPM mining depends combinatorially on the number of activities in the log, and therefore, mining LPMs on projections of the log instead of on the full significantly speeds up LPM mining. However, this results in a partial exploration of the LPM search space and does not guarantee that all LPMs meeting the support threshold are found. In principle, when the activities frequently following each other are in the same projection, the search space can be constrained almost without loss in quality of the mined LPMs. Such projection sets could potentially be overlapping. This is desired, since interesting patterns can potentially exist in some activity set $\{a, b, c\}$, as well as in $\{a, b, d\}$, and discovering on both $L\lceil_{\{a,b,c\}}$ and $L\lceil_{\{a,b,d\}}$ and then merging the results is faster than discovering on $\{a, b, c\} \cup \{a, b, d\} = \{a, b, c, d\}$. The typical approach to generate the projection set for LPM mining is to apply Markov graph clustering [32] to a graph where vertices represent activities and edges represent the *connectedness* of two activities $a$ and $b$ based

on following relations, i.e., $connectedness(a, b, L) = \sqrt{\dfrac{\#(\langle a,b\rangle, L)^2}{\#(\langle a\rangle, L)} + \dfrac{\#(\langle a,b\rangle, L)^2}{\#(\langle b\rangle, L)}}$ .

**Subtrace Mining** aims at finding frequent subsequences from logs. Diamantini et al. [9] apply *frequent subgraph mining* (FSM) to do so. In a first step, each trace $\sigma \in L$ is transformed into a directed graph $g = (V, E, \phi)$, with $V$ the set of nodes that correspond to events in $\sigma$, $E$ the set of the edges that show ordering relations between the events, and $\phi$ a labeling function associating nodes with the activities of the corresponding events. A node is created for each event in the trace and nodes representing subsequent events are connected with an edge. Once the set of graphs is obtained, an FSM algorithm is applied to derive frequent subgraphs from it, yielding the frequent subtraces in the event log. Diamantini et al. [9] use the SUBDUE algorithm [18] that adopts Description Length (DL) to iteratively select the most relevant subgraphs. Given a graph set $G$ and a subgraph $s$, SUBDUE uses an index based on DL, hereafter denoted by $\nu(s, G)$, which is computed as $\nu(s, G) = \frac{DL(G)}{DL(s) + DL(G|s)}$ where $DL(G)$ is the DL of $G$, $DL(s)$ is the DL of $s$ and $DL(G|s)$ is the DL of $G$ where each occurrence of $s$ in $G$ is replaced with a single node (i.e., compression). By doing so, SUBDUE relates the relevance of a subgraph with its compression capability.

At each iteration, it extracts the subgraph with the highest compression capability, i.e., the subgraph corresponding to the maximum value of the $\nu$ index. This subgraph is then used to compress the graph set. The compressed graphs are presented to SUBDUE again. These steps are repeated until no more compression is possible or until a user-defined number of iterations is reached. The outcome of SUBDUE consists of a set of subtraces ordered according to their relevance. As an example, Fig. 3 shows a portion of the SUBDUE output inferred from the set of graphs derived by the event log $L = \{\langle a, d, b, c, d, c, a, c, b, d, a, b, a\rangle^2, \langle a, e, c, c, d, c, a, c, b, d, a, b, a\rangle, \langle a, e, c, b, d\rangle\}$.

At the top level, we have subgraphs involving only elements of the original graphs set; while in the lower levels, we have subgraphs that involve upper level subgraphs in their definition. Since top-level subgraphs correspond to the most relevant subgraphs for the graphs set, we can reasonably expect to be able to capture most of the process behaviors by only considering the latter. Note that in this work we consider totally ordered traces, from which we construct sequential graphs; hence, the mined subgraphs are limited to sequential traces. However, SUBDUE can mine subgraphs that are more complex that just sequential traces, e.g., when partially ordered logs are analyzed.

## 3   Mining LPMs Using Subtrace Constraints and Projections

In this section, we present an approach to mine Local Process Models (LPMs) by exploiting subtrace mining results. This extends traditional LPM mining [33] by converting the set of subtraces (mined using the approach described in Sect. 2.2) into a set of *projections* (i.e., sets of activities) and a set of *ordering constraints* that are both used to restrict the set of possible expansions in the expansion phase of LPM mining. Projections restrict the possible extensions to the set of

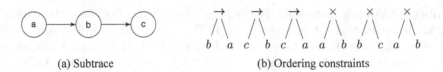

(a) Subtrace                    (b) Ordering constraints

**Fig. 4.** An example of subtrace and the corresponding ordering constraints.

activities in the projection, while ordering constraints prohibit expansions into LPMs that violate ordering relations.

Each subtrace represents a frequent and connected portion of the process. Activities that do not co-occur together in a subtrace are unlikely to co-occur in a frequent LPM. Therefore, we extract one projection for each subtrace, consisting of the activities in the subtrace. Furthermore, LPMs that directly contradict the behavior of a subtrace can be extracted as ordering constraints, as they unlikely represent execution orders between activities. Two types of constraints can be derived from subtraces: constraints on *exclusive choices* and constraints on *sequential executions*. Given a subgraph $s = (V, E, \phi)$, an ordering constraint on $s$ is defined as a process tree $M = \oplus(a, b)$ with $a, b \in V$ and $\oplus \in \{\rightarrow, \times\}$.

Algorithm 1 describes the procedure to convert a subgraph into a set of ordering constraints. The algorithm generates a constraint for the exclusive choice tree between each pair of activities in the subtrace and adds it to the set of ordering constraints (line 4). Then, the algorithm checks for each pair of vertices in the subtrace whether there exists a path (possibly transitively) from vertex $v_i$ to $v_j$ (line 5). If so, a sequential constraint is added, thus prohibiting the reversed ordering (line 6). Figure 4 shows the set of ordering constraints extracted for an example subtrace. Figure 4a shows that most occurrences of activity $a$ occurred before $b$, which, in turn, mostly occurred before $c$. Transitively, this means that $a$ occurred before $c$. The three leftmost trees in Fig. 4b show the extracted ordering constraints that directly contradict those frequent orderings in the subtrace. Furthermore, the existence of the subtrace indicates that the activities tend to co-occur and do not tend to be mutually exclusive. Therefore, we can safely remove from the LPM search space the process trees that contain exclusive choices constructs between those activities.

Algorithm 2 describes how to mine LPMs from a log $L$ given a set of subtraces $S$. For each subtrace in $S$, its set of activities and the ordering constraints are extracted (line 3), yielding set of projections and constraints $CP$. Algorithm *MiningLPMwithProjectionsAndConstraints* is invoked on the event log and the mined set of projections and constraints (line 13). This procedure mines LPMs in the traditional way, with an additional step in which every generated expansion $M_i \in Exp(M)$ of LPM $M$ is first checked against the set of ordering constraints $OC$. If there exists a constraint $oc \in OC$ such that $oc$ is a subtree of $M_i$, then $M_i$ is discarded and not further expanded.

---

**Algorithm 1.** Method *FindOrderingConstraints*

---

**Input** : subtrace $s = (V, E, \phi)$
**Output:** set of ordering constraints $OC$

1   $OC = \{\}$;
2   **foreach** $v_i \in V$ **do**
3      **foreach** $v_j \in V \setminus \{v_i\}$ **do**
4         $OC = OC \cup \{\times(\phi(v_i), \phi(v_j))\}$;
5         **if** *existsPath*$(v_i, v_j, E)$ **then**
6           $OC = OC \cup \{\rightarrow (\phi(v_j), \phi(v_i))\}$;
7   **return** $OC$;

---

**Algorithm 2.** Mining LPM using Subtraces

---

**Input** : event log $L$, set of subtraces $S$
**Output:** set of local process models $LPM$

1   $CP = \langle \rangle$;
2   **foreach** $s_i = (V_i, E_i, \phi_i) \in S$ **do**
3      $CP = CP \cdot \langle (\{\phi_i(v_i)|v_i \in V_i\}, findOrderingConstraints(s_i)) \rangle$;
4   $CP' = \emptyset$;
5   **foreach** $i \in \{1, 2, \ldots, |CP|\}$ **do**
6      $(V_i, OC_i) = CP(i)$;
7      **if** $\exists j \in \{1, 2, \ldots, |CP|\} : V_i \subset V_j \vee (V_i = V_j \wedge j > i)$ **then**
8         *continue*;
9      **foreach** $j \in \{1, 2, \ldots, i - 1\}$ **do**
10        **if** $(V_i \subseteq V_j \vee V_j \subseteq V_i)$ **then**
11          $OC_i = OC_i \cup OC_j$;
12      $CP' = CP' \cup \{(V_i, OC_i)\}$
13   **return** *MiningLPMwithProjectionsAndConstraints*$(L, CP')$;

---

## 4   Deriving Partial Order Relations over LPMs

In this section, we present an approach to discover partially ordered sets of Local Process Models (LPMs), which we will refer to as PO-LPMs. We adopt the approach in [13] for the mining of partial order relations between subtraces and adapt it to mine such relations between LPMs. We extract the following *ordering relations* between pairs of LPMs:

(i)   the *sequential* relation, denoted as $LPM_1 \rightarrow_{seq} LPM_2$, indicates that $LPM_2$ occurs immediately after $LPM_1$;
(ii)   the *concurrent* relation, denoted as $LPM_1 \rightarrow_{conc} LPM_2$, indicates that the executions of the activities in the LPMs are interleaved;
(iii)   the *eventually* relation, denoted as $LPM_1 \rightarrow_{ev} LPM_2$, indicates that $LPM_2$ occurs after $LPM_1$, but at least one other activity occurs between the two LPMs.

Given a set of LPMs *LPMS* and log $L$, the approach first reduces *LPMS* by *removing redundant ones* and then builds an *occurrence matrix* indicating

| $\sigma_1:$ $\langle$ a b c f l m g o n r $\rangle$ | | $\sigma_2:$ $\langle$ a b c f g a b c n r $\rangle$ | |
|---|---|---|---|

| $\sigma_1:$ | $\langle$ a b c f l m g o n r $\rangle$ |
|---|---|
| $LPM_1^1$ | × × × |
| $LPM_2^1$ | ⠀⠀⠀⠀× ⠀× |
| $LPM_3^1$ | ⠀⠀⠀⠀⠀⠀× × ⠀× × |

| $\sigma_2:$ | $\langle$ a b c f g a b c n r $\rangle$ |
|---|---|
| $LPM_1^1$ | × × × |
| $LPM_1^2$ | ⠀⠀⠀⠀⠀⠀× × × |
| $LPM_2^1$ | ⠀⠀⠀⠀× × |
| $LPM_3^1$ | |

| | $LPM_1^1$ | $LPM_1^2$ | $LPM_2^1$ | $LPM_3^1$ |
|---|---|---|---|---|
| $\sigma_1$ | 1 | 0 | 1 | 1 |
| $\sigma_2$ | 1 | 1 | 1 | 0 |

(a)⠀⠀⠀⠀⠀⠀⠀⠀⠀⠀⠀⠀(b)⠀⠀⠀⠀⠀⠀⠀⠀⠀⠀⠀⠀(c)

**Fig. 5.** Building of the occurrence matrix for $LPM_1$, $LPM_2$ and $LPM_3$ and traces $\sigma_1$, $\sigma_2$.

in which traces each LPM occurs. Finally, it derives sets of the LPMs that frequently co-occur by applying frequent itemset mining to the occurrence matrix and then extracts the PO-LPM for each itemset by inferring the ordering relation on the log for each pair of LPMs in the itemset. We now explain each step in more detail.

*Redundancy Reduction:* First we apply existing techniques to remove *redundant* LPMs from the mined set of LPMs, i.e. LPMs that only describe behavior that is already represented by other LPMs in the set. This simplifies and speeds up the partial orders inferring step. We use the redundancy reduction technique of [30], which uses a greedy search approach to find a subset of LPMs that maximizes the number of events in the log covered while minimizing the number of LPMs used.

*Occurrence Matrix:* We build an occurrence matrix $OM$ for event log $L$ and the LPMs $LPMS'$ obtained using redundancy reduction, where each cell $c_{ij}$ represents whether the $j$-$th$ LPM occurs in the $i$-$th$ trace. We build $OM$ using segmentation function $\Gamma$. As shown in Sect. 2.2, function $\Gamma$ can identify multiple instances of *the same LPM* in a single trace, therefore, in theory, multiple ordering relations can hold for a given pair of LPMs on a given trace. To deal with this property of $\Gamma$, we consider multiple instances of an LPM in a trace as *different* LPMs. Whenever we have more than one instance in a trace, we create a copy of the LPM for each of its occurrences and we set corresponding cells in the matrix to 1.

For example, consider the set of LPMs consisting of $LPM_1 = \{\rightarrow (\rightarrow (a, b), c)\}$, $LPM_2 = \{\wedge(f, g)\}$ and $LPM_3 = \{\wedge((\rightarrow (l, m), (\rightarrow (o, n))\}$. Figure 5 shows the occurrence matrix for these LPMs on traces $\sigma_1$ (Fig. 5a) and $\sigma_2$ (Fig. 5b), marking the events in the trace that belong to each LPM with ×. All three LPMs occur exactly once in $\sigma_1$, resulting in "1" values for all three LPMs in the occurrence matrix of Fig. 5c. In contrast, $\sigma_2$ contains *two* instances of $LPM_1$ (i.e., $LPM_1^1$ and $LPM_1^2$) and one of $LPM_2$.

*Deriving PO-LPMs:* We infer sets of LPMs that frequently co-occur in the same trace by applying any *frequent itemset mining* algorithm (see [12] for an

|        | $LPM_1$ | $LPM_2$ | $LPM_3$ |
|--------|---------|---------|---------|
| $LPM_1$ | 2 | 174 | 0 |
| $LPM_2$ | 0 | 0 | 0 |
| $LPM_3$ | 0 | 0 | 0 |

$M_{seq}$

|        | $LPM_1$ | $LPM_2$ | $LPM_3$ |
|--------|---------|---------|---------|
| $LPM_1$ | 0 | 0 | 0 |
| $LPM_2$ | 0 | 0 | 199 |
| $LPM_3$ | 0 | 0 | 0 |

$M_{conc}$

|        | $LPM_1$ | $LPM_2$ | $LPM_3$ |
|--------|---------|---------|---------|
| $LPM_1$ | 0 | 25 | 199 |
| $LPM_2$ | 0 | 0 | 0 |
| $LPM_3$ | 0 | 0 | 0 |

$M_{ev}$

(a) Ordering relation matrices

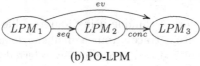

(b) PO-LPM

**Fig. 6.** Deriving PO-LPMs for itemset $\{LPM_1, LPM_2, LPM_3\}$.

overview) using a support threshold $\rho$. Then, we determine the ordering relations between the LPMs per set of frequently co-occurring LPMs. For each set of frequently co-occurring LPMs we extract the traces from the log in which these LPMs co-occur. Using $\Gamma$ we obtain the instances of the LPMs in these traces, from which we can extract the starting and ending position of each instance and determine whether $LPM_i \rightarrow_{seq} LPM_j$, $LPM_i \rightarrow_{conc} LPM_j$, or $LPM_i \rightarrow_{ev} LPM_j$ holds with $LPM_i, LPM_j$ two co-occurring LPMs. For each pair of LPMs occurring in the same itemset we store the number of traces for which these relations between the LPMs hold respectively in matrices $M_{seq}$, $M_{conc}$ and $M_{ev}$. Based on these matrices we extract as ordering relations between LPMs those relations that exceed a user-defined support threshold $\eta$, resulting in the PO-LPMs. Note that with $\rho$ and $\eta$ there are two distinct support thresholds. This is motivated by the fact that a pair of LPMs can occur in different order in different traces, and therefore the support of an ordering relation can be smaller than the support of the itemset. Note that $\eta$ can be considered as the *confidence* of the ordering relations.

As an example, consider again LPMs $LPM_1, LPM_2, LPM_3$ and trace $\sigma_1$, in which all three LPMs occur. Analyzing the positions of the events belonging to each LPM in Fig. 5a we observe that $LPM_2$ occurs immediately after $LPM_1$ (i.e., $LPM_1 \rightarrow_{seq} LPM_2$), that $LPM_2$ is interleaved with $LPM_3$ (i.e., $LPM_2 \rightarrow_{conc} LPM_3$) and $LPM_3$ eventually occurs after $LPM_1$ (i.e., $LPM_1 \rightarrow_{ev} LPM_3$). Suppose that for some set of LPMs and log $L$ that among others contains $\sigma_1$, itemset $\{LPM_1, LPM_2, LPM_3\}$ is extracted as a set of frequently co-occurring LPMs, and $M_{seq}$, $M_{conc}$ and $M_{ev}$ are as shown in Fig. 6a, then Fig. 6b shows the PO-LPM for this itemset for $\eta = 50\%$ of the traces. Note that the use of the thresholds $\rho$ and $\eta$ ensures us to infer PO-LPMs that meet minimum support requirements, as in the case of single LPMs.

# 5   Evaluation

In this section, we describe two sets of experiments. First, we evaluate the speedup in Local Process Model (LPM) mining that is obtained by applying the technique of Sect. 3. Secondly, we explore the resulting PO-LPMs obtained by applying the technique of Sect. 4. We evaluate both on the same collection of real-life event logs, which is described below.

*Datasets:* We evaluated our technique using four real-life event logs. The first event log contains execution traces from a financial loan application process at a large Dutch financial institution, commonly referred to as the *BPI'12* log [10]. This log consists of 13087 traces (loan applications) for which a total of 164506 events have been executed, divided over 23 activities. The second event log contains traces from the receipt phase of an environmental permit application process at a Dutch municipality, to which we will refer as the *receipt phase WABO* log [4]. The receipt phase WABO log contains 1434 traces, 8577 events, and 27 activities. The third event log contains medical care pathways of sepsis patients from a medium size hospital, to which we will refer as the *SEPSIS* log [24]. The SEPSIS log contains 1050 traces, 15214 events, and 16 activities. Finally, as fourth event log we use a dataset from the lighting system of a smart office environment, which was gathered in [36]. This dataset contains continuous values for the color temperature and the light intensity of the lighting in four different areas in the office space. Events correspond to interactions with the lighting interface that result in changes in the color temperature and intensity of the lighting in one or more areas and each case is a working day. The event names are converted from continuous values to symbolic activities using the well-known technique SAX [21], resulting in eight categories for each event representing the color temperature and intensity in each of the four areas. We refer to this log as *Laplace* and it contains 92 traces, 1557 events and 218 activities.

## 5.1   Mining LPMs Using Subtraces

We now explore the effect of using subtraces to the efficiency of LPM mining, for which we perform two sets of experiments. First, we investigate the effects of only using SUBDUE projections, i.e., using the projection-based LPM mining procedure of [32] while using the activities in SUBDUE subtraces as projections. Then, we exploit both projections and ordering constraints as described in Sect. 3.

*Tools and Configurations:* We use the iterative Markov LPM mining algorithm implemented in the *LocalProcessModelDiscovery* package[1] of the ProM framework [34]. We have implemented the novel LPM mining approach based on SUB-DUE projections and constraints in the ProM package *LocalProcessModelDiscoveryWithSubdueConstraints*[2]. For both Markov-based LPM and subtrace-based

---

[1] https://svn.win.tue.nl/repos/prom/Packages/LocalProcessModelDiscovery/.

[2] https://svn.win.tue.nl/repos/prom/Packages/LocalProcessModelDiscoveryWithSubdueConstraints.

LPM mining we use the standard ProM configurations. For SUBDUE we use the standard implementation[3], in which we varied the number of iterations. Note that a high number of SUBDUE iterations is expected to be beneficial for the quality of the LPM results: more iterations lead to a more process behavior being captured in subtraces. However, this negatively impacts the speedup of LPM mining. Moreover, by construction, SUBDUE extracts the largest frequent subtraces in the first iterations. Hence, we expect the obtained subtraces to be able to represent most of the process behaviors even using only a few iterations. Therefore, we explore using 1, 5 and 10 iterations, and to verify our assumption we additionally use 10000 iterations. All experiments are performed on an 2.4 GHz Intel i7 machine, equipped with 16 Giga of RAM.

*Methodology:* We evaluate our approach using two dimensions. First, we consider the reduction in the search space size, which represents how much speedup is obtained in the mining procedure. Secondly, we consider the quality of the mined LPMs by comparing the LPMs mined using SUBDUE projections and constraints to those mined when using the full search space. This second dimension is relevant since using projections with LPM mining might lead to not all LPMs being found [32]. By comparing the LPM rankings obtained by mining with and without projections we can assess to what extent the use of projections affects the results. We compare the ranking using *Normalized Discounted Cumulative Gain* (NDCG) [6,17], which is a widely used metrics to evaluate ranked results in information retrieval. Generally, NDCG@k is used, which only considers the top $k$ elements of the ranking. NDCG consists of two components, *Discounted Cumulative Gain* (DCG) and *Ideal Discounted Cumulative Gain* (IDCG). DCG aggregates the relevance scores (i.e., the score obtained with respect to the quality metrics) of individual LPMs in the ranking in such a way that the graded relevance is discounted with logarithmic proportion to their position in the ranking. This results in more weight being put on the top of the ranking compared to lower parts of the ranking. Formally, DCG is defined as: $DCG@k = \sum_{i=1}^{k} \frac{2^{rel_i}-1}{log_2(i+1)}$, where $rel_i$ is the relevance of the LPM at position $i$. Normalized Discounted Cumulative Gain (NDCG) is obtained by dividing the DCG value by the DCG on the ground truth ranking (called Ideal Discounted Cumulative Gain). Normalized Discounted Cumulative Gain (NDCG) is defined as: $NDCG@k = \frac{DCG@k}{IDCG@k}$.

As a baseline we apply the Markov-based projection technique from [32] iteratively until all projections contain at most seven activities, and compare the search space reduction and the NDCG obtained when using this approach with the search space reduction and NDCG obtained when using projections and constraints from SUBDUE subtraces.

---

[3] http://ailab.wsu.edu/subdue/.

**Table 1.** The search space size and NDCG results for mining LPMs with and without SUBDUE projections and constraints.

| Event Log | Projections (iterations) | Constraints (iterations) | Search space size | Speedup | NDCG@5 | NDCG@10 | NDCG@20 |
|---|---|---|---|---|---|---|---|
| BPI'12 | None | None | 1567250 | - | 1.0000 | 1.0000 | 1.0000 |
| | Iterative Markov | None | 36032 | 43.50x | 0.9993 | 0.9987 | 0.9865 |
| | Iterative Markov | SUBDUE (10k) | 21084 | 74.33x | 0.9993 | 0.9987 | 0.9865 |
| | SUBDUE (1) | None | 10608 | 147.74x | 0.9993 | 0.9987 | 0.9830 |
| | SUBDUE (5) | None | 10740 | 145.93x | **1.0000** | **0.9994** | 0.9870 |
| | SUBDUE (10) | None | 10904 | 143.73x | **1.0000** | **0.9994** | **0.9903** |
| | SUBDUE (10k) | None | 12666 | 123.74x | **1.0000** | **0.9994** | **0.9903** |
| | SUBDUE (1) | SUBDUE (1) | 2718 | 576.62x | 0.9993 | 0.9987 | 0.9830 |
| | SUBDUE (5) | SUBDUE (5) | **2620** | **598.19x** | **1.0000** | **0.9994** | 0.9870 |
| | SUBDUE (10) | SUBDUE (10) | 2874 | 545.32x | **1.0000** | **0.9994** | **0.9903** |
| | SUBDUE (10k) | SUBDUE (10k) | 4012 | 390.64x | **1.0000** | **0.9994** | **0.9903** |
| Receipt phase | None | None | 1451450 | - | 1.0000 | 1.0000 | 1.0000 |
| | Iterative Markov | None | 12074 | 120.21x | 0.9418 | 0.8986 | 0.8238 |
| | Iterative Markov | SUBDUE (10k) | 10610 | 136.80x | 0.9418 | 0.8986 | 0.8238 |
| | SUBDUE (1) | None | 8176 | 177.53x | **1.0000** | **0.9994** | 0.9903 |
| | SUBDUE (5) | None | 8256 | 175.81x | **1.0000** | **0.9994** | 0.9903 |
| | SUBDUE (10) | None | 8264 | 175.64x | **1.0000** | **0.9994** | **0.9958** |
| | SUBDUE (10k) | None | 8504 | 170.68x | **1.0000** | **0.9994** | **0.9958** |
| | SUBDUE (1) | SUBDUE (1) | 4012 | 390.64x | **1.0000** | **0.9994** | 0.9903 |
| | SUBDUE (5) | SUBDUE (5) | **1862** | **779.51x** | **1.0000** | **0.9994** | 0.9903 |
| | SUBDUE (10) | SUBDUE (10) | 2170 | 668.71x | **1.0000** | **0.9994** | **0.9958** |
| | SUBDUE (10k) | SUBDUE (10k) | 2178 | 666.41x | **1.0000** | **0.9994** | **0.9958** |
| SEPSIS | None | None | 315451 | - | 1.0000 | 1.0000 | 1.0000 |
| | Iterative Markov | None | 6304 | 50.04x | 0.9332 | 0.9148 | 0.8613 |
| | Iterative Markov | SUBDUE (10k) | 3768 | 83.72x | 0.9332 | 0.9148 | 0.8613 |
| | SUBDUE (1) | None | 12 | 26287.58x | 0.5763 | 0.3771 | 0.2489 |
| | SUBDUE (5) | None | 334 | 994.46x | 0.9916 | 0.9671 | 0.9472 |
| | SUBDUE (10) | None | 394 | 800.64x | **0.9923** | **0.9692** | **0.9534** |
| | SUBDUE (10k) | None | 1034 | 05.08x | **0.9923** | **0.9692** | **0.9534** |
| | SUBDUE (1) | SUBDUE (1) | **10** | **31545.10x** | 0.5763 | 0.3771 | 0.2489 |
| | SUBDUE (5) | SUBDUE (5) | 174 | 1812.94x | 0.9916 | 0.9671 | 0.9472 |
| | SUBDUE (10) | SUBDUE (10) | 144 | 2190.63x | **0.9923** | **0.9692** | **0.9534** |
| | SUBDUE (10k) | SUBDUE (10k) | 470 | 671.17x | **0.9923** | **0.9692** | **0.9534** |
| Laplace | None | None | 4784569 | - | 1.0000 | 1.0000 | 1.0000 |
| | Iterative Markov | None | 1096 | 4365.48x | 0.8261 | 0.7942 | 0.7635 |
| | Iterative Markov | SUBDUE (10k) | 746 | 6413.63x | 0.8261 | 0.7942 | 0.7635 |
| | SUBDUE (1) | None | 12 | 398714.08x | 0.4354 | 0.2836 | 0.1841 |
| | SUBDUE (5) | None | 42 | 113918.31x | 0.5061 | 0.3296 | 0.2139 |
| | SUBDUE (10) | None | 72 | 66542.35x | 0.7909 | 0.5683 | 0.3689 |
| | SUBDUE (10k) | None | 730 | 6554.20x | **0.9096** | **0.8690** | **0.7928** |
| | SUBDUE (1) | SUBDUE (1) | **10** | **478456.90x** | 0.4354 | 0.2836 | 0.1841 |
| | SUBDUE (5) | SUBDUE (5) | 34 | 140722.62x | 0.5061 | 0.3296 | 0.2139 |
| | SUBDUE (10) | SUBDUE (10) | 56 | 85438.73x | 0.7909 | 0.5683 | 0.3689 |
| | SUBDUE (10k) | SUBDUE (10k) | 582 | 8220.91x | **0.9096** | **0.8690** | **0.7928** |

*Results:* Table 1 shows the results for the four logs. The results obtained without using projections or constraints are considered to be the ground truth LPM ranking and therefore have NDCG@k values of 1.0 by definition. The results obtained by the best heuristic configuration(s) are reported in bold and between parenthesis is the number of SUBDUE iterations.

Iterative Markov [32] projections result in a reduction of the search space by a factor between 43.50x (BPI'12) and 4365.48x (Laplace), while the high NDCG values indicate that the majority of the top 20 LPMs of the ground truth are still found. Using the constraints extracted from SUBDUE subtraces obtained with 10k SUBDUE iterations together with iterative Markov projections further increases the speedup of LPM mining on all four logs while resulting in identical LPM rankings.

The search space size of LPM mining with SUBDUE projections depends on the number of iterations performed by SUBDUE: more iterations result in a larger number of unique sets of activities, leading to more projections and a larger LPM search space, but at the same time increasing the quality of the mined LPMs in terms of NDCG. Note that the quality of the LPM mining results differs between the logs when one SUBDUE iteration is used. This is because for logs with few activities a single subtrace can already capture most relevant process behavior, while for logs with many activities it can only capture a small part. The use of SUBDUE projections leads to a higher speedup than iterative Markov projections on all logs, even SUBDUE constraints are not used. At the same time, when enough SUBDUE iterations are used, SUBDUE projections result in higher NDCG. This shows that SUBDUE subtraces are more effective in finding related sets of activities for use as projections in LPM mining compared to Markov clustering.

The constraints extracted from SUBDUE subtraces in combination with SUBDUE-based projections results in considerably higher speedup on all logs without resulting in lower NDCG. On three logs, using 10 SUBDUE iterations is sufficient to achieve the highest quality LPMs, while only on the Laplace log more iterations are needed. Using SUBDUE projections and constraints we have found speedups between 598.19x (BPI'12) and 478456.90x (Laplace). To put these results into perspective: this brought down the mining time on the BPI'12 log from 24 min to less than two minutes. This shows that subtrace mining results can be used to speed up LPM mining. Additionally, in [31] we showed that the mined LPMs provide additional process insights in comparison to subtraces, meaning that it is actually useful to perform LPM mining after subtrace mining.

## 5.2   Mining Ordering Relations over LPMs

In this section, we evaluate our approach to discover PO-LPMs from a set of LPMs. We propose a set of measures to assess the quality of PO-LPMs and we discuss the results that we obtained for the four logs. Note that the notion of quality exploited in these experiments differs from the one used before. In the previous experiments, the quality of the different LPMs set was intended as their similarity with the set of LPMs discovered by the exhaustive search, to evaluate the impact of the pruning of the search state. Here, we focus on exploring the benefits of considering larger and possible *unconnected* portions of process behaviors. Therefore, we evaluate the balance between the loss in support and the gaining in size of PO-LPMs with respect to single LPMs sets.

Additionally, we show how PO-LPMs can be used to merge LPMs resulting in higher-level LPMs that describe a larger fragment of the process.

*Tools and Configurations:* For each log we use the set of LPMs that we obtained in the experiments of Sect. 5.1 for projections using 10k SUBDUE iterations and use the implementation of the technique to reduce redundancy in LPM results [30] as available in ProM package *LocalProcessModelConformance*[4]. We implemented the PO-LPM mining approach of Sect. 4 in PHP[5] and use the implementation of the FP-Growth itemset mining algorithm in the SPMF pattern mining library [11] to obtain the frequent itemsets (*FI* hereafter), i.e. sets of frequently co-occurring LPMs.

*Methodology:* We test our technique with three types of sets of *FI*: (1) the entire set of *FI*; (2) the set of *closed* FI, i.e. the subset of *FI* where for each itemset $i$ there exists no other itemset $j$ such that $i \subset j$ with identical support to $i$; (3) the set of *maximal* FI, i.e. *FI* where for each itemset $i$ there exists no other itemset $j$ with $i \subset j$ where the support of $j$ exceeds $\rho$. We vary $\rho$ from 1% to 100% increasing it in steps of 1%. We set $\eta = 50\%$ since, as a rule of thumb, it is reasonable to consider only ordering relations occurring at least in more than half of the cases in which the LPMs occur together. Lower values for $\eta$ would likely result in PO-LPMs involving multiple and infrequent relations between pairs of LPMs, thus affecting the understandability and the representative capability of the output. We evaluate the quality of the discovered PO-LPMs along two dimensions: (1) the amount of information provided on the process (i.e., pattern size) and (2) the portion of process behaviors they represent (i.e., their support). It is easy to see that there is a trade-off between these dimensions: larger patterns typically have lower support. What is the optimal trade-off between the two dimensions depends on the process analysis task at hand and needs to be decided by the process analyst. Here, we investigate the trade-off between the dimensions as a result of $\rho$. To capture both dimensions in a single measure we also define *Information Ratio* (IR) measure as follows: $IR = \frac{\#activitiesLPM}{\#activitiesProcess} \times \frac{\#occurrences}{\#traces}$. Function *IR* yields values in interval $[0, 1]$ with 0 corresponding to an empty set of LPMs and 1 corresponding to a set of LPMs that involves all process activities and occurs in all traces.

*Results:* Table 2 reports statistics on the set of PO-LPMs that are inferred from three of the four logs for the three different itemset mining approaches, as well as for the original set of LPMs ("-" in column *FI*). Columns *#LPMs, Avg. #Act, Avg. Supp (%)*, and *Avg. IR* respectively indicate the number of the LPMs in the set, the average number of activities per LPM, the average support of the LPMs, and the average information ratio. The receipt phase log is missing in the table, as only two LPMs remained after the redundancy reduction step, between which no ordering relation could be found.

---

[4] https://svn.win.tue.nl/repos/prom/Packages/LocalProcessModelConformance.
[5] https://surfdrive.surf.nl/files/index.php/s/PeD64m5xr5hxcqi.

**Table 2.** LPMs set statistics inferred from the three event logs.

| Log | FI | #LPMs | Avg. #Act | Avg. supp (%) | Avg. IR |
|-----|-----|-------|-----------|---------------|---------|
| BPI'12 | - | 5 | 2 | 41.6 | 0.036 |
| | All | 210 | 7.05 | 3.16 | 0.006 |
| | Closed | 9 | 7.6 | 20.5 | 0.035 |
| | Maximal | 9 | 7.6 | 20.5 | 0.035 |
| SEPSIS | - | 5 | 2.4 | 53.8 | 0.090 |
| | All | 136 | 7.61 | 3.7 | 0.015 |
| | Closed | 46 | 7.85 | 7.6 | 0.037 |
| | Maximal | 26 | 8.69 | 8.8 | 0.038 |
| Laplace | - | 18 | 2 | 9 | 0.0009 |
| | All | 65758 | 15.97 | 1 | 0.0007 |
| | Closed | 21 | 8.19 | 1 | 0.0004 |
| | Maximal | 21 | 8.19 | 1 | 0.0004 |

It should be noted that by combining the LPMs inferred from each log, for the BPI'12 log we derived 10 activities out of 23 activities in the process. Respectively for the SEPSIS and Laplace logs we derived 10 out of 16 and 30 activities out of 218 activities. This suggests that the processes under analysis involve many infrequent activities that do not need to be modeled to capture most of the structure in the process, and therefore, they are not in the LPM set. In turn, this implies that most of the LPMs involve only a small fraction of the process activities. This explains why the IR values overall are very small, regardless of log and settings. However, this does not affect our analysis, since we investigate how IR values vary on the same log between different PO-LPMs sets instead of considering their absolute value.

**BPI'12.** All *FI* configurations led to PO-LPMs involving over three times the amount of activities of the initial LPMs. Using all PO-LPMs results in a large support drop, while it does not lead to larger LPMs compared to the closed and maximal PO-LPMs. Figure 7a shows more detailed results by plotting support against the number of activities for single LPMs, and after merging them using all, closed, and maximal frequent itemsets. We discretized support values in bins of 5%. Each dot in the plot represents the set of LPMs that involve $n$ activities and has a support within $[s-0.05, s]$, where $s$ denotes the support represented by the bin. The larger the size of the dot is, the larger the size of the corresponding set of LPMs is.

All configurations led to LPMs with a dimension at least double than and up to six times the dimension of single LPMs. The set of all PO-LPMs involves a high number of LPMs with a support smaller or equal to 5%, which motivates the low support values and, in turn, the worsening of the IR values with respect to the single set. Note that most of these low-support PO-LPMs were discarded

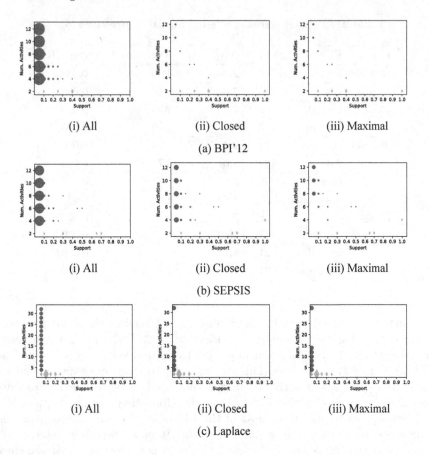

**Fig. 7.** The support and the number of activities of each LPM in the result set using for original LPMs (red diamonds) compared with PO-LPMs (blue circles) obtained using all, closed and maximal frequent itemsets. (Color figure online)

in the closed and maximal sets, which involve only 10 LPMs each, against the 210 of all PO-LPMs. However, most of the LPMs in these sets have a support lower than or equal to 20%, thus leading to an average support value equal to around the half of single LPMs.

Figure 8 presents an example of a merged LPM that is built from the PO-LPM obtained from the closed (maximal) set. The PO-LPM is represented as a Petri net (introduced in Sect. 2.1), where circles represent places, rectangles represent transitions, and black rectangles depict $\tau$-transitions. Places that belong to the initial marking contain a token and places belonging to a final marking are marked as ◎ The dotted lines surround the original LPMs and edges between original LPMs are labeled with their ordering relation. This merged LPM consists of eventually relations between LPM 1 and LPM 3 and between LPM 3 and LPM 2. This PO-LPM shows that after submitting a loan application it

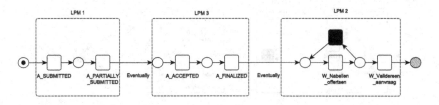

**Fig. 8.** One of the PO-LPMs from the closed set for BPI'12.

was accepted and finalized, followed by one or more calls to the customer for additional information and finally a validation of the application documents. This merged LPM occurs in 25% of the traces, which is significant given its size. Note that LPMs of this size cannot be mined with existing techniques. Given its size and support this PO-LPM provides the analyst with a higher-level and more meaningful representation of the process compared to the three LPMs separately.

**SEPSIS.** We obtained a small number of single LPMs, mostly comprising two activities, with one LPM involving 4 activities. PO-LPMs are on average three times larger than the single LPMs. However, for this log the increase in size was not enough to properly balance the loss in terms of support; indeed, all configurations achieved IR values worse than the set of original LPMs. The set of all PO-LPMs is again the one with the lowest *IR* value, while the closed and maximal sets have similar performance. Figure 7b provides the scatter plot of size/support for LPMs obtained from the SEPSIS log for all tested configurations. The PO-LPMs have a size up to six times the one of most single LPMs; however, many of them have a support between 1% and 5%. Some of these low-support LPMs were not filtered neither in the closed nor in the maximal set.

Figure 9 reports one of the PO-LPMs with the highest support. It starts with the registration of the patient in the emergency room (ER), followed by filling the general triage document (*ER Triage*), which is done concurrently to either filling in a triage form for sepsis cases (*ER SEPSIS Triage*) or the infusion of some liquids (*IV Liquid*). Later in the process, LPM 4 and LPM 5 are executed in parallel. LPM 4 shows that the patient was admitted into the normal care ward and CRP was performed (i.e., a test to detect inflammation); LPM 5 shows that the patient's leukocytes were tested and she was then sent back to the emergency room. Also here, we obtained a meaningful description of interconnected phases of the process and obtained a reasonable support value (i.e., 12%).

**Laplace.** Both single LPMs and PO-LPMs lead to very low IR values for the Laplace log. The reason is that this log is much less structured than the other logs we analyzed, resulting in LPMs with a low support (1%–2%). Moreover, because this process contains many activities, these LPMs involve a small fraction of those activities. On this log, differently from the other logs, the all itemsets configuration obtains high IR values than the others. All configurations led

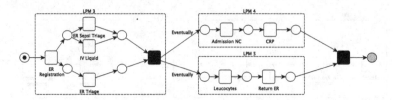

**Fig. 9.** One of the PO-LPMs from the maximal set for SEPSIS.

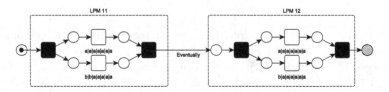

**Fig. 10.** One of the PO-LPMs from the closed set for Laplace.

to an average support of 1%; however, PO-LPMs obtained from all frequent itemsets have twice the average size compared to closed and maximal itemsets. The main reason is that the maximal and closed sets filter out many large (but non-maximal or non-closed) PO-LPMs. Figure 7c provides the scatter plots for size/support for LPMs in the Laplace log for all tested configurations. As expected, we got many large LPMs in the lowest support interval for all configurations. This is particularly evident for the first configuration. It is worth noting that PO-LPMs were able to achieve growth of up to 15 times the average size compared to single LPMs, although this growth comes with a decrease of support from around 40% to a maximum of 1%.

Figure 10 shows one of the closed PO-LPMs with the highest support mined for the Laplace log. The figure shows that when we have a couple of signals, occurring in every order, representing a switch either from low to high values or from high to low values for both the color tone and the light intensity for the first office area, this pair is eventually followed by another pair of signals, representing a switch of the light intensity values in the same area. The support of this PO-LPM is 2%.

## 6    Related Work

We discuss two areas of related work: *subprocess mining* and *partial order discovery*.

*Subprocess Mining.* Several approaches have been proposed to extract the most relevant subprocesses (intended as subgraphs) from a set of process execution traces. Some approaches propose to extract subprocesses from sequential traces [3,16]. For instance, Bose et al. [3] mine subprocesses by identifying sequences of events that fit a priori defined templates. Compared to these approaches, our approach does not require any predefined template and extracts subprocesses that are the most relevant according to their description length.

Several other techniques, like LPM mining, focus on the mining of more complex patterns that allow for control-flow constructs. Chapela-Campa et al. [8] developed a technique called WoMine-i to mine subprocess patterns with multiple control-flow constructs that are *infrequent*. Lu et al. [22] recently proposed an interactive subprocess exploration tool, which allows the discovery of subprocess patterns that a process analyst can modify based on domain knowledge. Greco et al. [14] propose a Frequent Subgraph Mining (FSM) algorithm that exploits knowledge about relationships among activities (e.g., AND/OR splits) to drive subgraphs mining. Graphs are generated by replaying traces over the process model; however, this algorithm requires a model properly representing the event log, which may not be available for many real-world processes.

*Partial Order Discovery.* The discovering of partial ordering relations among log events has been traditionally addressed by *Episode Discovery* [25]. An episode is defined as a collection of partially ordered events. The goal of Episode Discovery consists in determining all the episodes in an event log whose support is above a user-defined threshold. Episodes are usually detected by grouping together events falling in the same window (e.g., a time or a proximity window), generating all possible candidates (i.e., all possible partial orders configuration) and then checking the frequency of the candidates. Since the seminal work of Mannila et al. [25], several approaches have been proposed to enhance the efficiency of episode discovery, addressing different application domains (e.g., [15,37]). Recently, Leemans et al. [19] introduced an approach tailored to discover episodes from event logs generated by business processes, where it is possible to exploit the notion of process instance to determine the episodes. The output of their approach consists of directed graphs where nodes correspond to activities and edges to eventually follow precedence relations. Our work presents some similarities with [19], in the sense that also the discovery of our PO-LPMs is based on the notion of process instances. However, our work defines ordering relations among patterns of events, rather than between single events. Moreover, our approach provides a more fine grained analysis by distinguishing among sequential, eventually and concurrency relations.

# 7    Conclusions and Future Work

In this work, we have explored the synergy effects between subtrace and LPM mining, showing how their combination enables the gathering of relevant process insights that would remain hidden when both are applied separately. Specifically, we extended the LPM algorithm in [32] to account for ordering constraints mined using SUBDUE subtraces. Moreover, we proposed an approach (adapting the approach of [13]) to derive ordering relations between LPMs to infer partial orders between them. We evaluated our approach on four real-world event logs. The results show that mining LPMs with SUBDUE projections and constraints outperforms the current state-of-the-art techniques for LPM mining both in quality as well as in computation time. Our experiments also show that the approach is able to infer partially ordered models, thereby providing a more complete and meaningful overview on the process compared to single LPMs, although this comes at the price of a loss in support.

In future work, we plan to explore the use of other subtrace mining techniques to derive LPMs. Moreover, we plan to investigate semi-automatic techniques to move from PO-LPMs to process models expressed in a standard notation by converting the partial relations in actual process constructs. This will allow the reuse of the discovered process patterns for further analysis.

**Acknowledgement.** This work is partially supported by ITEA3 through the APP-STACLE project (15017) and by the RSA-B project SeCludE.

# References

1. van der Aalst, W.M.P.: Process Mining: Data Science in Action. Springer, Heidelberg (2016). https://doi.org/10.1007/978-3-662-49851-4
2. van der Aalst, W.M.P., Adriansyah, A., van Dongen, B.F.: Replaying history on process models for conformance checking and performance analysis. Wiley Interdiscip. Rev.: Data Min. Knowl. Discov. **2**(2), 182–192 (2012)
3. Jagadeesh Chandra Bose, R.P., van der Aalst, W.M.P.: Abstractions in process mining: a taxonomy of patterns. In: Dayal, U., Eder, J., Koehler, J., Reijers, H.A. (eds.) BPM 2009. LNCS, vol. 5701, pp. 159–175. Springer, Heidelberg (2009). https://doi.org/10.1007/978-3-642-03848-8_12
4. Buijs, J.C.A.M.: Receipt phase of an environmental permit application process ('WABO'). CoSeLoG project (2014). https://doi.org/10.4121/uuid:a07386a5-7be3-4367-9535-70bc9e77dbe6
5. Buijs, J.C.A.M., van Dongen, B.F., van der Aalst, W.M.P.: A genetic algorithm for discovering process trees. In: CEC, pp. 1–8. IEEE (2012)
6. Burges, C., et al.: Learning to rank using gradient descent. In: ICML, pp. 89–96 (2005)
7. Carmona, J., Cortadella, J., Kishinevsky, M.: A region-based algorithm for discovering petri nets from event logs. In: Dumas, M., Reichert, M., Shan, M.-C. (eds.) BPM 2008. LNCS, vol. 5240, pp. 358–373. Springer, Heidelberg (2008). https://doi.org/10.1007/978-3-540-85758-7_26
8. Chapela-Campa, D., Mucientes, M., Lama, M.: Discovering infrequent behavioral patterns in process models. In: Carmona, J., Engels, G., Kumar, A. (eds.) BPM 2017. LNCS, vol. 10445, pp. 324–340. Springer, Cham (2017). https://doi.org/10.1007/978-3-319-65000-5_19
9. Diamantini, C., Genga, L., Potena, D.: Behavioral process mining for unstructured processes. J. Intell. Inf. Syst. **47**(1), 5–32 (2016)
10. van Dongen, B.F.: BPI challenge (2012). https://doi.org/10.4121/uuid:3926db30-f712-4394-aebc-75976070e91f
11. Fournier-Viger, P., Gomariz, A., Gueniche, T., Soltani, A., Wu, C.W., Tseng, V.S.: SPMF: a Java open-source pattern mining library. J. Mach. Learn. Res. **15**(1), 3389–3393 (2014)
12. Fournier-Viger, P., Lin, J.C.W., Vo, B., Chi, T.T., Zhang, J., Le, H.B.: A survey of itemset mining. Wiley Interdiscip. Rev.: Data Min. Knowl. Discov. **7**(4) (2017)

13. Genga, L., Potena, D., Martino, O., Alizadeh, M., Diamantini, C., Zannone, N.: Subgraph mining for anomalous pattern discovery in event logs. In: Appice, A., Ceci, M., Loglisci, C., Masciari, E., Raś, Z.W. (eds.) NFMCP 2016. LNCS (LNAI), vol. 10312, pp. 181–197. Springer, Cham (2017). https://doi.org/10.1007/978-3-319-61461-8_12
14. Greco, G., Guzzo, A., Manco, G., Saccà, D.: Mining and reasoning on workflows. IEEE Trans. Knowl. Data Eng. **17**(4), 519–534 (2005)
15. Huang, K.Y., Chang, C.H.: Efficient mining of frequent episodes from complex sequences. Inf. Syst. **33**(1), 96–114 (2008)
16. Huang, Z., Lu, X., Duan, H.: On mining clinical pathway patterns from medical behaviors. Artif. Intell. Med. **56**(1), 35–50 (2012)
17. Järvelin, K., Kekäläinen, J.: Cumulated gain-based evaluation of IR techniques. ACM Trans. Inf. Syst. **20**(4), 422–446 (2002)
18. Jonyer, I., Cook, D., Holder, L.: Graph-based hierarchical conceptual clustering. J. Mach. Learn. Res. **2**, 19–43 (2002)
19. Leemans, M., van der Aalst, W.M.P.: Discovery of frequent episodes in event logs. In: Ceravolo, P., Russo, B., Accorsi, R. (eds.) SIMPDA 2014. LNBIP, vol. 237, pp. 1–31. Springer, Cham (2015). https://doi.org/10.1007/978-3-319-27243-6_1
20. Leemans, S.J.J., Fahland, D., van der Aalst, W.M.P.: Discovering block-structured process models from event logs containing infrequent behaviour. In: Lohmann, N., Song, M., Wohed, P. (eds.) BPM 2013. LNBIP, vol. 171, pp. 66–78. Springer, Cham (2014). https://doi.org/10.1007/978-3-319-06257-0_6
21. Lin, J., Keogh, E., Lonardi, S., Chiu, B.: A symbolic representation of time series, with implications for streaming algorithms. In: SIGMOD Workshop on Research Issues in DM&KD, pp. 2–11. ACM (2003)
22. Lu, X., et al.: Semi-supervised log pattern detection and exploration using event concurrence and contextual information. In: Panetto, H., et al. (eds.) CoopIS. LNCS, vol. 10573. Springer, Heidelberg (2018). https://doi.org/10.1007/978-3-319-69462-7_11
23. Maggi, F.M., Mooij, A.J., van der Aalst, W.M.P.: User-guided discovery of declarative process models. In: CIDM, pp. 192–199. IEEE (2011)
24. Mannhardt, F., Blinde, D.: Analyzing the trajectories of patients with sepsis using process mining. In: RADAR+EMISA, pp. 72–80. CEUR (2017)
25. Mannila, H., Toivonen, H., Verkamo, A.I.: Discovery of frequent episodes in event sequences. Data Min. Knowl. Discov. **1**(3), 259–289 (1997)
26. Mǎruşter, L., van Beest, N.R.T.P.: Redesigning business processes: a methodology based on simulation and process mining techniques. Knowl. Inf. Syst. **21**(3), 267–297 (2009)
27. Ramezani, E., Fahland, D., van der Aalst, W.M.P.: Where did i misbehave? Diagnostic information in compliance checking. In: Barros, A., Gal, A., Kindler, E. (eds.) BPM 2012. LNCS, vol. 7481, pp. 262–278. Springer, Heidelberg (2012). https://doi.org/10.1007/978-3-642-32885-5_21
28. Reisig, W.: Petri Nets: An Introduction, vol. 4. Springer, Heidelberg (2012). https://doi.org/10.1007/978-3-642-69968-9
29. Schönig, S., Cabanillas, C., Jablonski, S., Mendling, J.: Mining the organisational perspective in agile business processes. In: Gaaloul, K., Schmidt, R., Nurcan, S., Guerreiro, S., Ma, Q. (eds.) CAISE 2015. LNBIP, vol. 214, pp. 37–52. Springer, Cham (2015). https://doi.org/10.1007/978-3-319-19237-6_3
30. Tax, N., Dumas, M.: Mining non-redundant sets of generalizing patterns from sequence databases. arXiv preprint arXiv:1712.04159 (2017)

31. Tax, N., Genga, L., Zannone, N.: On the use of hierarchical subtrace mining for efficient local process model mining. In: Proceedings of International Symposium on Data-driven Process Discovery and Analysis, pp. 8–22. CEUR-WS.org (2017)
32. Tax, N., Sidorova, N., van der Aalst, W.M.P., Haakma, R.: Heuristic approaches for generating local process models through log projections. In: CIDM, pp. 1–8. IEEE (2016)
33. Tax, N., Sidorova, N., Haakma, R., van der Aalst, W.M.P.: Mining local process models. J. Innov. Digit. Ecosyst. **3**(2), 183–196 (2016)
34. Verbeek, H.M.W., Buijs, J.C.A., Van Dongen, B.F., van der Aalst, W.M.P.: ProM 6: the process mining toolkit. In: BPM Demos, vol. 615, pp. 34–39. CEUR (2010)
35. van der Werf, J.M.E.M., van Dongen, B.F., Hurkens, C.A.J., Serebrenik, A.: Process discovery using integer linear programming. In: van Hee, K.M., Valk, R. (eds.) PETRI NETS 2008. LNCS, vol. 5062, pp. 368–387. Springer, Heidelberg (2008). https://doi.org/10.1007/978-3-540-68746-7_24
36. van de Werff, T., Niemantsverdriet, K., van Essen, H., Eggen, B.: Evaluating interface characteristics for shared lighting systems in the office environment. In: DIS, pp. 209–220. ACM (2017)
37. Zhou, W., Liu, H., Cheng, H.: Mining closed episodes from event sequences efficiently. In: Zaki, M.J., Yu, J.X., Ravindran, B., Pudi, V. (eds.) PAKDD 2010. LNCS (LNAI), vol. 6118, pp. 310–318. Springer, Heidelberg (2010). https://doi.org/10.1007/978-3-642-13657-3_34

# A Linear Temporal Logic Model Checking Method over Finite Words with Correlated Transition Attributes

Jean-Michel Couvreur[1] and Joaquín Ezpeleta[2(✉)]

[1] Laboratoire d'Informatique Fondamental d'Orléans (LIFO), Université d'Orléans,
Orléans, France
jean-michel.couvreur@univ-orleans.fr
[2] Department of Computer Science and Systems Engineering,
Aragón Institute of Engineering Research (I3A),
University of Zaragoza, Zaragoza, Spain
ezpeleta@unizar.es

**Abstract.** Temporal logic model checking techniques are applied, in a natural way, to the analysis of the set of finite traces composing a system log. The specific nature of such traces helps in adapting traditional techniques in order to extend their analysis capabilities. The paper presents an adaption of the classical Timed Propositional Temporal Logic to the case of finite words and considers relations among different attributes corresponding to different events. The introduced approach allows the use of general relations between event attributes by means of freeze quantifiers as well as future and past temporal operators. The paper also presents a decision procedure, as well as a study of its computational complexity.

**Keywords:** Model checking · Freeze Linear Temporal Logic ·
Conformance checking · Log analysis

## 1 Introduction

Current information systems usually generate log files to record system and user activities. System logs contain very valuable information that, when properly analyzed, could help in getting a better understanding of the system and user behaviors and then in improving the system. In many (most) cases the log can be seen as a set of *traces*: a trace is a chronologically ordered sequence of events corresponding to a process execution. It can correspond, for instance, to the events of a user session in an e-commerce website or database, the events corresponding to the execution of a process in a workflow system, etc.

This work was done when J. Ezpeleta was a visiting researcher at the University of Orléans. It has been partially supported by the TIN2017-84796-C2-2-R project, granted by the Spanish Ministry of Economy, Industry and Competitiveness.

© IFIP International Federation for Information Processing 2019
Published by Springer Nature Switzerland AG 2019
P. Ceravolo et al. (Eds.): SIMPDA 2017, LNBIP 340, pp. 89–104, 2019.
https://doi.org/10.1007/978-3-030-11638-5_5

Process mining [1] is the set of techniques that try to analyze log files looking for trace patterns so as to synthesize a model representing the set of event sequences in the log. In some cases, when the system is governed by a rather closed procedural approach, the model itself can be a quite constraining and closed model (Petri net, BPMN process, etc.) fitting the log. In cases in which the system does not really constrain the user possibilities (open or turbulent environments), the utility of such constraining models decreases: the log can contain so many different combinations of event sequences that the obtained model will be a kind of *spaghetti* or *flower*, depending on the variety of such combinations. For the last cases, a viable approach consists of establishing a set of behavioral properties (described in a high level formalism, such as temporal logic, for instance) describing possible model behaviors and then checking which traces in the log satisfy them, looking for what are usually called a *declarative process* (described in an implicit way by the set of formulas).

Conformance checking is the process by which a log and a model (either a procedural model or a declarative one) are compared, so as to get a measure of how well the log and the model are aligned. This paper concentrates on the conformance perspective using a variant of temporal logic for property description and a model checker for conformance checking. Temporal logic has been extensively used in process mining [20,22]. Initially, only control flow aspects were considered. A case (trace) was defined as an ordered sequence of activities. Later, a multi-perspective point of view was adopted. In this case, each event, besides the activity, could contain additional data, as the time at which the event happened, the resource that executed the activity, the duration, etc. [9,17,19,21]. As shown in [17], conformance results can significantly vary when data associated to activities is considered.

Classical LTL temporal logic for declarative process conformance imposes some constraints with respect to the kind of properties one can deal with. Let us consider, as an example, a trace whose events are of the form $(ac, re, ts)$, where $ac$ stands for the activity, $re$ for the resource that executed it and $ts$ for the event time-stamp. It is possible to express by means of a classical LTL formula the property that a concrete activity $a$ executed by a concrete resource $r$ is always ($G$ temporal logic operator) followed by a future ($F$ temporal logic operator) concrete activity $b$ executed by the same concrete resource $r$: $G((a, r) \rightarrow F(b, r))$. However, it is not possible to express the property that a concrete activity $a$ is followed in the future by activity $b$, and both activities are carried out by the same resource (being the resource "any" resource). In the case of finite domains one could transform such formula into the disjunction of a set of formulas (one per resource). However, this is infeasible for general data domains. Consider, for instance, the necessity of correlating the times at which the considered events happened, so as to ensure that both are in an interval of 30 min.

Focusing on real-time applications, different extensions of LTL have been proposed in the literature with the aim of incorporating time and time-related event correlations. Metric Temporal Logic (MTL) [13] considers the until modality with an interval window of validity. Timed Propositional Temporal Logic

(TPTL) [2] adds *freeze* variables as the way of referring and correlating to specific time values associated to different word positions. Metric First Order Temporal Logic (MFOTL) [4] extends MTL with first order quantifiers, gaining in description power. In the domain of log analysis, the work in [6] gives a big step forward towards the full integration of the control and data perspectives, considering the time as a special part of the data associated with events. Authors use MFOTL for the specification of behavioral properties, and propose model checking functions for a subset of MP-Declare [21] patterns.

Freeze-like operators have also been applied in specific application domains. [3] defines Biological Oscillators Synchronization Logic (BOSL) for the specification and verification of global synchronization properties for a set of coupled oscillators. [5] defines the STL* logic (extended Signal Temporal Logic) for checking temporal properties on continuous signals representing behaviors of biological systems.

In this paper we propose the DLTL temporal logic as an adaption and extension of TPTL which allows a real integration of the data, control and time perspectives from both, future and past perspectives. The way the logic is defined allows working with any data attributes associated to events as well as general relations among them. The approach can be considered as an integrated multi-perspective conformance checking method. The main contributions in the paper are: (1) the proposal of the DLTL temporal logic able to deal with a whole multi-perspective point of view; (2) the proposal of a general model checking algorithm for such logic, with no constrain about the set of formulas that can be analyzed and (3) the space and time complexity characterization of the proposed model checking method.

The paper is organized as follows. Section 2 formally defines the logic and also describes it by means of some intuitive examples. Section 3 proposes a model checking algorithm and evaluates its time and space complexity. Section 4 shows how the proposed logic and model checking are applied to the analysis of a log corresponding to a workflow system, used in the literature. Section 5 briefly describes a model checker prototype. Section 6 comments on some related work which concentrate on (timed) temporal logic and model checking approaches. Finally Sect. 7 establishes some conclusions of the work and gives some future perspectives for its continuation.

## 2   DLTL

The logic we are proposing is based on the the Timed Propositional Temporal Logic, TPTL, [2]. TPTL is a very elegant formalism which extends classical linear temporal logic with a special form of quantification, named *freeze quantification*. Every freeze quantifier is bound to the time of a particular state. As an example, the property "whenever there is a request $p$, and variable $x$ is frozen to the current state, the request is followed by a response $q$ at time $y$, so that $y$ is at most, $x+10$" is expressed in TPTL by the formula $Gx.(p \rightarrow Fy.(q \wedge (y \leq x + 10)))$ [2]. Since the formula requires to talk about two different points in the trace ($p$ and

$q$ states), two freeze variables are used in order to be able to correlate the time values of those states, and also the required constraint that both instants must verify: $x$ and $y$ instants must not be separated more than 10 time units. A TPTL formula can contain as many freeze operators as required.

The adaption of TPTL that we propose focuses on two different aspects. On the one hand, we generalize the kind of relations between the attributes of the events corresponding to freeze operators. TPTL constrained event correlations to checking equality and usual relational operation between the attribute values of freeze variables (positions in the word). In DLTL event correlations are allowed to be more general (as general as any function correlating any attribute values). On the second hand, DLTL also incorporates past temporal operators. Without them, some interesting properties relating current and past word positions could not be expressed.[1]

Freezing a variable by means of a freeze variable $x$ will allow us to talk about attributes of the event at that position, and then establish correlations between attributes of different events by means of relations, of the form "The resource associated to $x$ is different than the resource associated to $y$" or "The price of such product doubles between events separated more than two days", for instance. The timestamp of an event can be considered as just another attribute. In the case of timestamp attributes we are going to assume they are coherent with the ordering of events in the trace, so that if event $e_1$ appears before than event $e_2$, the timestamp of $e_1$ will be no greater than the one of $e_2$ (the trace is monotonic with respect to such attribute).

Let us now formally introduce the DLTL logic.

**Definition 1.** *Let $\mathcal{D}$ be a set, called the* transition domain; *let $\mathcal{V} = \{x_1, x_2, \ldots\}$ be a finite set of* freeze variables *and let $\Phi = \{\varphi_1(x_1^1, \ldots, x_{m_1}^1), \varphi_2(x_1^2, \ldots, x_{m_2}^2), \cdots \mid m_i \geq 0, \ x_j^i \in \mathcal{V}, \ \forall i, j\}$ be a finite set of* relations *on $\mathcal{D}$.*

*The set of correct formulas, $\mathcal{F}_{(\mathcal{D}, \mathcal{V}, \Phi)}$, for the DLTL logic, is inductively defined as follows:*

- *$f \in \Phi$ is a correct formula*
- *If $x \in \mathcal{V}$ and $f_1$, $f_2$ are correct formulas, so are $\neg f_1$, $f_1 \wedge f_2$, $X f_1$, $Y f_1$, $f_1 \ U \ f_2$, $f_1 \ S \ f_2$, $x.f_1$*

In the previous definition, a relation with one variable will be called a *proposition*.

**Definition 2.** *Let $f \in \mathcal{F}_{(\mathcal{D}, \mathcal{V}, \Phi)}$ be a correct DLTL formula. A* valuation $v$ *is a mapping from the set of variables in $f$ into $\mathcal{D}$.*

DLTL formulas of $\mathcal{F}_{(\mathcal{D}, \mathcal{V}, \Phi)}$ will be interpreted over non-empty finite words of elements of $\mathcal{D}$, of the form $\sigma = \sigma_1 \cdot \sigma_2 \cdot \ldots \cdot \sigma_n$ (as usual $|\sigma|$ denotes the

---

[1] In the original logic, atomic formulas where associated to states. Since we are going to concentrate on log traces, the point of view we adopt associates general data to events.

length of the word). In order to make notations simpler, in the following, for a giving word $\sigma$, when talking about a valuation $v$ we will assume that for any variable $x$, $v(x)$ is one of the sets in the word, identified by its position in $\sigma$ and, therefore, $1 \leq v(x) \leq n$.

Let us now define when a correct formula is satisfied by a word at a given transition:

**Definition 3.** Let $f \in \mathcal{F}_{(\mathcal{D}, \mathcal{V}, \Phi)}$ be a correct DLTL formula; let $\sigma = \sigma_1 \cdot \sigma_2 \cdot \ldots \cdot \sigma_n$ be a finite word over $\mathcal{D}$; let $v$ be a valuation and let $i$ be an index such that $1 \leq i \leq n$. By $\sigma, i \models_v f$ we denote that $\sigma$ satisfies $f$ for valuation $v$ at position $i$. This relation is defined as follows:

- $\sigma, i \models_v \top$
- $\sigma, i \models_v p$ if $p(\sigma_i)$, for any proposition $p$
- $\sigma, i \models_v \varphi(x_1, \ldots, x_m)$ if $\varphi(\sigma_{v(x_1)}, \ldots, \sigma_{v(x_m)})$.
- $\sigma, i \models_v \neg f$ if $\neg(\sigma, i \models_v f)$
- $\sigma, i \models_v f_1 \wedge f_2$, for any pair $f_1$ and $f_2$, if $\sigma, i \models_v f_1$ and $\sigma, i \models_v f_2$
- $\sigma, i \models_v Xf$, for any formula $f$, if $i < n$ and $\sigma, i+1 \models_v f$
- $\sigma, i \models_v Yf$, for any formula $f$, if $1 < i$ and $\sigma, i-1 \models_v f$
- $\sigma, i \models_v f_1 \ U \ f_2$, for any pair $f_1$ and $f_2$, if there exists an index $i \leq k \leq n$ such that $\sigma, k \models_v f_2$ and, for any $i \leq j < k$, $\sigma, j \models_v f_1$
- $\sigma, i \models_v f_1 \ S \ f_2$, for any pair $f_1$ and $f_2$, if there exists an index $j \leq i$ such that $\sigma, j \models_v f_2$ and, for any $j+1 \leq k \leq i$, $\sigma, k \models_v f_1$
- $\sigma, i \models_v x.f$, for any formula $f$ and variable $x$ if $\sigma, i \models_{v[x \leftarrow i]} f$, where $v[x \leftarrow i]$ represents the valuation such that $v[x \leftarrow i](x) = i$ and $v[x \leftarrow i](y) = v(y)$ for any $y \neq x$.

In the formula $x.f$, $f$ is the scope of the freeze variable $x$. To avoid misinterpretations, we are not allowing to rebind a variable inside its scope. The set of operators is extended with the classical abbreviations: $f_1 \vee f_2 \equiv \neg(\neg f_1 \wedge \neg f_2)$, $Ff \equiv \top U f$, $Gf \equiv \neg(F \neg f)$, $f \Rightarrow g \equiv g \vee \neg f$, $f \Leftrightarrow g \equiv (f \Rightarrow g) \wedge (g \Rightarrow g)$, $O f \equiv \top S f$, $Hf \equiv \neg(O \neg f)$ (here $O$ operator stands for *Once*), and $\bot = \neg \top$.

*Example 1.* As a first example, let us consider a trace of the execution of a process. Let us consider a set of agents, $Ag = \{a, b, c\}$, a set of actions, $Ac = \{req, ack, other\}$, and let $\mathcal{D} = Ag \times Ac \times \mathbb{R}$. Let us now consider the following word, corresponding to a trace of length 5:

$$\sigma = (a, req, 2)(b, req, 4)(a, ack, 6)(c, other, 8)(b, ack, 13)$$

For short, given $d \in \mathcal{D}$, $d.ag$, $d.act$ and $d.t$ will denote the first, second and third components, respectively.

The property that *for any agent, every req is followed by the corresponding ack of the same agent within a given time interval of 8 time units* can be expressed in DLTL this property can be established as follows:

$$f_1 = G(x.(\varphi_1(x) \Rightarrow Fy.(\varphi_2(x, y) \wedge \varphi_3(x) \wedge \varphi_4(x, y))))$$

with $\varphi_1(x) = (x.act = req)$, $\varphi_2(x,y) = (x.ag = y.ag)$, $\varphi_3(x) = (x.act = ack)$ and $\varphi_4(x,y) = (y.t - x.t \leq 8)$, being, in this case, $\Phi = \{\varphi_1, \varphi_2, \varphi_3, \varphi_4\}$.

In this example, variable $x$ is used to "freeze" a position in the word, while variable $y$ refers to a later position. $\varphi_3$ and $\varphi_4$ establish two different relations among the attributes in that positions.

*Example 2.* Considering the same example, we can also easily express the property that *every ack must be preceded by a req of the same agent in the previous 8 time units*

$$f_2 = G(x.((x.act = ack) \Rightarrow O(y.((x.ag = y.ag) \wedge (y.act = req) \wedge (x.t - y.t \leq 8)))))$$

*Example 3.* Let us now consider two sets, $A$ and $B$, with characteristics functions $\mathcal{C}_A$ and $\mathcal{C}_B$, respectively. And let us assume we want to state the property that every pair of positions $x$ and $y$ verify the relation $\varphi(x,y)$. This property could be checked by means of the following formula:

$$G(x.(\mathcal{C}_A(x) \Rightarrow H(y.(\mathcal{C}_B(y) \Rightarrow \varphi(x,y)))) \wedge G(y.(\mathcal{C}_B(y) \Rightarrow \varphi(x,y)))))$$

*Example 4.* Let us now assume that the third component corresponds to the event timestamp. The following formula expresses whether the *trace duration is greater than 10 time units*, which is true ($\neg X\top$ is true only at the last event):

$$x.(F(y.(\neg X\top \wedge (y.t - x.t > 10))))$$

*Example 5.* One interesting aspect is the possibility of referring to the position of an event in the trace, if we consider that each event has such position as an attribute. The following formula expresses whether the *trace contains at least 20 events*, which is false (# is the event attribute with its position inside the trace):

$$F(x.(\neg X\top \wedge (x.\# \geq 20)))$$

## 3 The Complexity of Model Checking a DLTL Formula

[12] presents a deep and clear study of the complexity of the problem of verifying a TPTL formula against a finite word, which can be easily adapted to the case of DLTL formulas. In this section we introduce a detailed description of the problem in DLTL with the aim of pointing out the reasons behind the cost of the verification process. Besides of proving that it is in PSPACE [12], we prove that it is exponential in time with respect to the number of freeze variables, and linear with respect to the rest of the involved parameters (size of the formula and length of the word).

We first introduce a recursive procedure for model checking DLTL formulas, and then we evaluate the complexity of the method. Since the complexity depends on the evaluation of the relations in $\Phi$ we are going to assume that the cost of evaluating such relations is "reasonable".

Checking function $dltl\_sat(\sigma, i, v, f)$ takes as parameters a word, $\sigma$, a position in the word, $i$, a valuation $v$ and a DLTL formula, $f$. Checking $f$ on the word $\sigma$ is carried out by means of the evaluation of $dltl\_sat(\sigma, 1, \emptyset, f)$ (valuation $v$ will be dynamically defined as long as the formula is checked). The algorithm is, basically, a recursive implementation of the inductive definition of DLTL formulas. In the case of parameter $i$ being outside the range of $\sigma$, we consider the formula is false. Freeze variables are considered as word position variables. In the case of $f$ being a relation $\varphi(x_1, \ldots, x_m)$, we assume in the evaluation of $dltl\_sat(\sigma, i, v, f)$ that valuation $v$ binds a value for each variable $x_1, \ldots, x_m$. This way evaluating the function is the same as evaluating $\varphi(v(\sigma_{x_1}), \ldots, v(\sigma_{x_m}))$. Notice that such evaluation does not depend on parameter $i$ (provided $i$ is in the range of $\sigma$). Evaluating a formula $x.f$ for a position $i$ is the same a making $x = i$ in $v$.

```
function dltl_sat(σ,i,v,f)
    if i ≤ 0 or i > |σ| then
        return false
    elseif f = φ(x₁,...,xₘ) then
        return φ(v(σ_{x₁}),...,v(σ_{xₘ}))
    elseif f = p then
        return p(σ_i)
    elseif f = Xf₁ then
        return dltl_sat(σ,i + 1,v,f₁)
    elseif f = f₁Uf₂ then
        return dltl_sat(σ,i,v,f₂) ∨ (dltl_sat(σ,i,v,f₁) and dltl_sat(σ,i + 1,v,f))
    elseif f = Yf₁ then
        return dltl_sat(σ,i − 1,v,f₁)
    elseif f = f₁Sf₂ then
        return dltl_sat(σ,i,v,f₂) ∨ (dltl_sat(σ,i,v,f₁) and dltl_sat(σ,i − 1,v,f))
    elseif f = x.f₁ then
        local old_x = v[x]
        v[x] = i
        ans = dltl_sat(σ,i,v,f₁)
        v[x] = old_x
        return ans
end
```

Let us now concentrate on the complexity of the proposed algorithm. The cost clearly depends on the cost of evaluating relations $\varphi(v(\sigma_{x_1}), \ldots, v(\sigma_{x_m}))$. We are going to assume that they are PSPACE with respect to the size of $f$ (as usual, the size is the number of operands and operators in the formula) and the length of $\sigma$, $|\sigma|$. With respect to the time, we are going to denote $K_{|f|,|\sigma|}$ a bound for all of them.

The following propositions establish the time and space complexity of $dltl\_sat(\sigma, i, v, f)$.

**Proposition 1.** *The model checking problem for $\sigma \models f$, where $\sigma$ is a finite word and $f$ is a DLTL formula, is PSPACE.*

*Proof.* Evaluating $dltl\_sat(\sigma, 1, \emptyset, f)$ will require, at most, $|f|$ recursive invocations. At each invocation $dltl\_sat(\sigma, i, v, g)$, where $g$ is a subformula of $f$, valuation $v$ can be passed as a reference to an $|Var_f|$-indexed array, being $Var_f$ the set of freeze variables in the formula. On the other hand, $f$ is coded by its syntax tree being each subformula $g$ a node. As a consequences, the size of the parameters of each invocation are of constant size (the considered references plus the size of $old\_x$ when needed). Provided that we are assuming that evaluating $\varphi(v(\sigma_{x_1}), \ldots, v(\sigma_{x_m}))$ is PSPACE, we can conclude that evaluating $dltl\_sat(\sigma, 1, \emptyset, f)$ is also PSPACE.

In order to obtain a better time execution cost, we use dynamic programming techniques as the way of avoiding recomputing the same subformula more than once for the same parameters.

**Proposition 2.** *The model checking problem for $\sigma \models f$, where $\sigma$ is a finite word and $f$ is a DLTL formula, can be solved in $O((K_{|f|,|\sigma|} + |Var_f|) \times |\sigma|^{|Var_f|} \times |f| \times |\sigma|)$ time.*

*Proof.* Provided that the same subformula is not going to be computed more than once, $|\sigma| \times |\sigma|^{|Var_f|} \times |f|$ is an upper bound for the number of invocations. In the case the subformula is $\varphi(v(\sigma_{x_1}), \ldots, v(\sigma_{x_m}))$, the cost is $K_{|f|,|\sigma|}$. When out of the word range, the cost is constant. We have also to consider the cost added by the dynamic programming technique. For that, we can use an array of size $|Var_f| + 2$, so that the cost of looking for a value is $O(|Var_f|)$. This way, we can conclude.

Let us now prove that the problem of checking a DLTL formula for a finite word is PSPACE HARD. Let us first prove that the problem of satisfying a QBF (Quantified Boolean Formula) can be translated into a checking problem.

**Lemma 1.** *Let $\phi(x_1, \ldots, x_{2n})$ be a boolean formula. Let $\Phi$ the the following quantified boolean formula:*

$$\Phi = \forall x_1, \exists x_2, \ldots \forall x_{2n-1}, \exists x_{2n}, \phi(x_1, \ldots, x_{2n})$$

*Let us consider the word $\sigma = (1, true) \cdot (2, false) \ldots (2 * i + 1, true) \cdot (2i + 2, false) \ldots (2 * n - 1, true) \cdot (2n, false)$ and the following DLTL formula $f = y_0 \cdot L_\forall(1)$ defined as follows (for each transition $x$ in the word, $x.t$ and $x.val$ denote the first and second components, respectively):*

$$L_\forall(2n+1) = \phi(y_1.val, \ldots, y_{2n}.val)$$
$$L_\forall(i) = G(y_i \cdot (y_i.t - y_{i-1}.t \leq 1 \Rightarrow L_\exists(i+1)))$$
$$L_\exists(i) = F(y_i \cdot (y_i.t - y_{i-1}.t \leq 1 \wedge L_\forall(i+1)))$$

*Then $\Phi$ is true iff $\sigma$ fulfills the DLTL formula $f$.*

*Proof.* We are going to prove, by induction, that

$$L\Phi(2i+1)(y_1.t, \ldots, y_{2i}.t) = \forall x_{2i+1}, \exists x_{2i+2}, \ldots \forall x_{2n-1},$$
$$\exists x_{2n}, \phi(y_1.val, \ldots, y_{2i}.val, x_{2i+1}, \ldots x_{2n})$$

When $i = n$, $L_\forall(2n+1)$ does not depend on the position in the word, and the equality is verified everywhere:

$$L_\forall(2n+1) = \Phi(2n+1) = \phi(y_1.val, \ldots, y_{2n}.val)$$

Assuming now the property is satisfied for $i+1$, let us prove that it is also true for $i$. $L_\forall(2i+1)$ can be expressed in terms of $L_\forall(2i+3)$ as follows:

$$L_\exists(2i+2) = F(y_{2i+2} \cdot (y_{2i+2}.t - y_{2i+1}.t \leq 1 \wedge L_\forall(2i+3)))$$
$$L_\forall(2i+1) = G(y_{2i+1} \cdot (y_{2i+1}.t - y_{2i}.t \leq 1 \Rightarrow L_\exists(2i+2)))$$

Applying induction hypothesis for $L_\exists(2i+2)$ we get:

$$L_\exists(2i+2) = F(y_{2i+2} \cdot (y_{2i+2}.t - y_{2i+1}.t \leq 1 \wedge \Phi(2i+3)))$$

for every position until $2i+2$. When evaluating $F$ and freezing variable $y_{2i+2}$, only two non-trivial positions have to be considered: either the same position or the next one. Since two consecutive positions cover both boolean values, the formula can be simplified as follows:

$$L_\exists(2i+2) = \Phi(2i+3)(y_1, \ldots, y_{2i+1}, false) \wedge \Phi(2i+1)(y_1, \ldots, y_{2i+1}, true)$$
$$= \exists x_{2i+2}, \Phi(2i+3)(y_1, \ldots, y_{2i+1}, x_{2i+2})$$

Doing analogously for the formula $L_\forall(2i+1)$ and positions until $2i+1$, we reach the searched result:

$$L_\forall(2i+1) = \forall x_{2i+1} \exists x_{2i+2}, \Phi(2i+3)(y_1, \ldots, x_{2i+1}, x_{2i+2})$$
$$= \Phi(2i+1)$$

**Proposition 3.** *The model checking problem for $\sigma \models f$, where $\sigma$ is a finite word and $f$ is a DLTL formula, is PSPACE-Hard.*

*Proof.* Immediate from Lemma 1

## 4   An Application Example

As an application case, let us consider the log described and analyzed in [16][2]. The log corresponds to the trajectories, obtained from the merging of data from the ERP of a Dutch hospital, followed by 1050 patients admitted to the emergency ward, presenting symptoms of a sepsis problem. The total number of events was 15214. Each event is composed of the *activity* (there are 16 different activities, categorized as either medical or logistical activities -*ER Sepsis Triage, IV Antibiotics, LacticAcid, IV Liquid,...*-), as well as additional information (time-stamps, in seconds, of the beginning and end of the activities, data

---

[2] https://doi.org/10.4121/uuid:d9769f3d-0ab0-4fb8-803b-0d1120ffcf54.

from laboratory tests and triage checklists, etc.). [16] applies different automatic process discovery techniques and obtain different models.

The objective of this section is to show how some of the system requirements, including time constraints, can be expressed in terms of DLTL formulas and checked for conformance with the log. In the following, for an event $x$ of a trace, $x.a$ and $x.t$ correspond to the activity and time-stamp in seconds, respectively.

**Requirement:** "Between *ER Sepsis Triage* and *IV Antibiotics* actions should be less than $1\,h$"[3]. Since a-priori we do not know whether there exists any causal relation between the considered activities, we are going to check the requirement as follows. Let

$$r1\_0 = F(\text{ER Sepsis Triage}) \ \wedge \ F(\text{IV Antibiotics})$$
$$r1\_1 = F \ x.(\text{ER Sepsis Triage} \ \wedge$$
$$F \ y.(\text{IV Antibiotics} \ \wedge \ y.t - x.t \leq 3600))$$
$$r1\_2 = F \ x.(\text{IV Antibiotics} \ \wedge$$
$$F \ y.(\text{ER Sepsis Triage} \ \wedge \ y.t - x.t \leq 3600))$$

be the formulas that check how many traces contain both activities, how many execute the second activity no later than one hour after the first and how many execute the first activity no later than one hour after the second one, respectively. Checking $r1\_0$, $r1\_1$ and $r1\_2$ gives, respectively, 823, 342 and 0 positive answers. This means that the requirement is fulfilled in 41.5% cases and, therefore, violated in 58.5% of the cases. Notice also that there is a causal relation between both events, since the *ER Sepsis Triage* always precedes *IV Antibiotics*. This result coincides with the one presented in [16].

**Requirement:** "Between *ER Sepsis Triage* and *LacticAcid* should be less than $3\,h$". Let us now consider the following formulas:

$$r2\_0 = F(\text{ER Sepsis Triage})$$
$$r2\_1 = F(\text{ER Sepsis Triage}) \ \wedge \ F(\text{LacticAcid})$$
$$r2\_2 = F \ x.(\text{ER Sepsis Triage} \ \wedge$$
$$(F \ y.(\text{LacticAcid} \wedge (y.t - x.t \leq 10800)) \ \vee$$
$$O \ z.(\text{LacticAcid} \wedge (x.t - z.t \leq 10800))))$$

$r2\_0$ gives that there are 1048 cases in which *ER Sepsis Triage* happens, $r2\_1$ is satisfied by 859 cases while $r2\_2$ states there are 842 cases with the appropriate time distance between the considered events. If one just considers those cases in $r2\_1$, the property is held in 98.02%, and violated in only 1.98%. This result is different than the 0.7% reported in [16]. The discrepancy could be explained in the way requirements have been checked. In our case, we directly work with

---

[3] As in [16], we are using "$\leq$" to check the properties, besides "should be less than $1\,h$" suggests "$<$" should be used.

the log, considering every trace. However, [16] checks the requirement against a model extracted from the log using Multi-perspective Process Explorer, which fits 98.3% of traces. On the other hand, if one considers the time constraint must be verified for every case in which *ER Sepsis Triage* occurs ($r2\_0$), the property is true in only 80.34% of the cases.

**Requirement:** Another proposed question is related to the patients returning to the service. Formula $r3\_0 = F(\text{Return ER})$ gives 28% of positive answers (27.8% in [16]). They are also interested in knowing how many of them return within 28 days. This can be checked with the formula

$$r3\_1 = x.(F\ y.(\text{Return ER}\ \wedge\ y.t - x.t \le 28 * 24 * 3600))$$

obtaining 94 traces, a 8.95% (12.6% in [16]).

As an additional question, one could ask whether there is a relation between the two first requirements and the third one. As an example of a formula with more than two variables, let us check this property by means of the formula $r3\_1 \wedge r4$, where

$$
\begin{aligned}
r4 = F(x.(&\text{ER Sepsis Triage} \wedge \\
&(F\ y.(\text{IV Antibiotics} \wedge (y.t - x.t <= 3600)))\ \wedge \\
&(F\ z.(\text{LacticAcid} \wedge (z.t - x.t \le 10800)) \vee \\
&O\ w.(\text{LacticAcid} \wedge (x.t - w.t \le 10800)))))
\end{aligned}
$$

The result is 27 traces (out of 94), which means that only 28.7% of those patients that return within 28 days correspond to patients that verify the constraints of one and three hours previously checked, which can be pointing to the adequacy of respecting the established time intervals.

## 5  About the Model Checking Process

In this section, we briefly describe a way of implementing a DLTL model checker. The algorithm here described is different from the direct recursive description used in Sect. 3. Having the same complexity, the use of symbolic storage of formulas together with some techniques of dynamic programming allowed us to obtain better execution performances with this second approach.

In order to describe the way DLTL formulas can be checked, let us consider Example 1 again:

$$f = G(x.((x.act = req) \Rightarrow Fy.((x.ag = y.ag) \wedge (y.act = ack) \wedge (y.t - x.t \le 8))))$$

Walking over the syntax tree of the formula allows to build the tableau used for checking it, as in Table 1. After analyzing the leaves of the $\wedge$ subtrees, a row is added, whose column values are the symbolic representation of the formula

$\phi(x,y) = (x.ag = y.ag) \wedge (y.act = ack) \wedge (y.t - x.t \leq 8)$. Next, the $y.\phi(x,y)$ is evaluated: for each column $c$, $y$ must take the value $\sigma_c$, giving the corresponding $\phi(x,c)$ symbolic column. For instance $\phi(x,5) = (x.ag = b) \wedge (y.act = ack) \wedge (13 - x.t \leq 8)$. Next row corresponds to $F(y.\phi(x,y))$, and so on until the complete tree is evaluated. As a result, a vector of *true/false* values is obtained. The value in position 1 is the result of checking the formula for the word. In this case, the answer of the model checker is (and should be) *false*.

**Table 1.** Checking tableau for the formula in Example 1

| i | 1 | 2 | 3 | 4 | 5 |
|---|---|---|---|---|---|
| $\sigma_i$ | (a,req,2) | (b,req,4) | (a,ack,6) | (c,other,8) | (b,ack,13) |
| ... | ... | ... | ... | ... | ... |
| $\phi(x,y)$ | $\phi(x,y)$ | $\phi(x,y)$ | $\phi(x,y)$ | $\phi(x,y)$ | $\phi(x,y)$ |
| $f_1(x) = y \cdot \phi(x,y)$ | $\phi(x,1)$ | $\phi(x,2)$ | $\phi(x,3)$ | $\phi(x,4)$ | $\phi(x,5)$ |
| $f_2(x) = F(f_1(x))$ | $\exists i \geq 1, \phi(x,i)$ | $\exists i \geq 2, \phi(x,i)$ | $\exists i \geq 3, \phi(x,i)$ | $\exists i \geq 4, \phi(x,i)$ | $\phi(x,5)$ |
| $f_3(x) = (x.act = req)$ | $x.act = req$ | $x.act = req$ | $x.act = req$ | $x.act = req$ | $x.act = req$ |
| $f_4 = x \cdot (f_3(x) \Rightarrow f_2(x))$ | $f_3(1) \Rightarrow f_2(1)$ | $f_3(2) \Rightarrow f_2(2)$ | $f_3(3) \Rightarrow f_2(3)$ | $f_3(4) \Rightarrow f_2(4)$ | $f_3(5) \Rightarrow f_2(5)$ |
|  | True | False | True | True | True |
| $G(f_4)$ | False | False | True | True | True |

As stated, the required time can be exponential with respect to the number of freeze variables. In order to get insight of the real time required we have carried out some experiments measuring the user time required for checking formulas with an increasing number of freeze variables. For that, we have considered the following parametrized formula

$$\phi(n) = Fx_1.(Gx_2.(Fx_3.(Gx_4.(\ldots Fx_{2n-1}.(Gx_{2n}.(\bigwedge_{i=2}^{2n}(x.t_i - x.t_{i-1} \leq 100)\ldots)$$

Figure 1 shows the chart corresponding to checking the formula for different values of parameter $n$ against the sepsis log used in Sect. 4. The curve is as expected. It fits the exponential $y = 0.9270899856 \cdot e^{0.1030458522 \cdot x}$ ($R^2 = 0.9956898464$, $rss = 297.8737955$). Notice that the method is able to efficiently deal with "many" freeze variables. If we constraint ourselves to a set of usual patterns involving a small set of freeze variables (as it is the case of the DECLARE formalism, for instance, whose patterns require two freeze variables at most) the model checking method is quite efficient (for instance, checking $r2\_2$, which only involves three freeze variables, against the sepsis log, needed $0.07$ s).

The experiments have been carried out with a prototype of the model checker implemented in lua 5.3, and executed in a Intel(R) Core(TM) i7-4790K CPU @ 4.00 GHz computer with a Ubuntu 16.04 operating system.

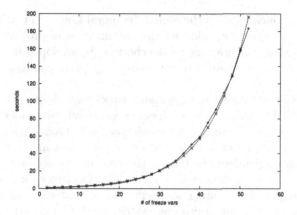

**Fig. 1.** Time versus number of freeze variables for formula $\phi(n)$, compared to $y = 0.9270899856 \cdot e^{0.1030458522 \cdot x}$ (continuous line corresponds to experimental results)

## 6 Related Work

Temporal logic with data has been used in different domains. [11] proposes Quantified-Free First-Order LTL (QFLTL($\mathbb{R}$)) where transitions, besides atomic propositions, can also contain real data attributes. QFLTL formulas are allowed to include classical operators between real expressions. Global variables can be used to correlate data of different transitions. Checking a formula is translated into finding intervals of the involved real variables verifying the constraints in the formula. The logic is constrained to some specific event structure and operations, of interest for the concrete domain it is proposed for. Temporal databases, together with temporal logic, have been used as a way to correlate time and data, allowing to analyze data correlations between the values of the database states at different time instants [7,8].

The addition of freeze variables (also named as *counters* in the domain) to classical LTL, as proposed in TPTL [2] allows correlating values of different points in a word. For the case of more general data in transitions (the term *dataword* is also used in the literature to refer to general words whose elements are data of a given domain), *freezeLTL* [10] is able to deal with correlations between attributes checking the equality of the considered values for a subset of TPTL. For the full TPTL, [12] studies the complexity of model-checking a TPTL formula against a finite word.

As stated in the introduction, freeze variables have been used in specific application domains. The Biological Oscillators Synchronization Logic, BOSL, introduced in [3], uses freeze operators for the specification of global synchronization properties for a set of coupled oscillators (modeled as a set of timed automata). Allowed propositions in the logic are constrained by the application domain and are comparisons of linear combinations of remaining times of oscillators at different time instants. The proposed model checking is a direct implementation of the recursive definition of the logical operators. [5] defines the STL* logic, which extends the Signal Temporal Logic, STL [15], adding the signal-value freeze operator, allowing the specification of properties related to damped oscillations. The way the model checking is developed imposes propositions in states to be constrained to comparisons of linear combinations of signal variables.

In the domain of process mining many works have dealt with conformance checking using LTL as the way of specifying behavioral properties. Since in most cases authors are interested in imposing or finding some process structures, they usually concentrate on a restricted set of patterns which reflect usual and interesting event dependencies. This is the case of the set of patterns in the *Declare* [18] workflow management system. The Declare approach focused on the control perspective, defining a specific set of patterns. Instances of such patterns define specific constrains the system must verify. [21] proposed MP-Declare, an extension of Declare including the data perspective of events. The paper also proposes a checking method for the considered logic, based on SQL.

[14] uses Timed-Declare as the formalism to add time to Declare. They constraint Metric Temporal Logic (MTL) [13] to the set of Declare patterns and adapt it to finite traces. Besides detecting that a constraint has already been violated, the proposed method can be used for the monitoring of the system evolution allowing an early detection that a certain constraint would be violated in the future, allowing for an a-priori guidance to avoid undesired situations.

In [6] the authors propose an approach which allows a multi-perspective point of view in which data and timestamps (those must be natural numbers) of events are considered as two parallel structures (according to [7], they adopt a snapshot perspective). MFOTL [4] (adapted for finite traces) is used as the formalism for the specification of properties. The paper reformulates MP-Declare patterns as MFOTL formulas, and presents a general framework for conformance checking. The framework is based on a general skeleton algorithm, which requires a different instance for each MP-Declare pattern.

The two previous methods, as stated, concentrate on a subset of MP-Declare, and specific methods must be developed for specific patterns, either as a specific function in the second case or as a specific SQL query in the first. On the other hand, given the specific application domain both methods are devoted, the proposed methods do not provide with a general procedure to model check any formula. The focus is on relations between pairs of events (the *activation* event, which imposes requirement conditions for the *target* event by means of a relation that must be satisfied by the corresponding associated data).

## 7   Conclusions

The paper has introduced a linear temporal logic able to deal with correlations among different values associated to different points in a finite word. Also, a model checking procedure has been introduced, and its complexity established in terms of the formula and word sizes. The interest of working with finite words comes from the fact the logic is going to be applied to the analysis of system logs.

For testing purposes, a model checker prototype has been developed in lua (not described in the paper) which has been used for the application example. The introduced method is general in the sense that it imposes no constraint neither with respect to the set of temporal logic formulas that can be checked nor with respect to the attributes that can be handled by the logic.

The interest of using the proposed approach is not limited to the case of turbulent environments, where process mining methods would generate spaghetti or flower models, but also in those cases in which a good model can be synthesized. The model itself can suggest implicit behavioral properties that could be model-checked against the log.

One direction for future work is to explore whether the proposed approach can be effectively used for complex logs with complex formulas. In the experiments we have carried out the response time was really short, but deeper analysis is necessary to deduce its applicability to big logs. The problem of dealing with a big number of traces can be alleviated by parallelizing the checking procedure: just use different parallel processors for dealing with different subsets of traces. The expensive dimensions are the length of the trace and the number of freeze variables.

# References

1. Agrawal, R., Gunopulos, D., Leymann, F.: Mining process models from work-flow logs. In: Schek, H.-J., Alonso, G., Saltor, F., Ramos, I. (eds.) EDBT 1998. LNCS, vol. 1377, pp. 467–483. Springer, Heidelberg (1998). https://doi.org/10.1007/BFb0101003
2. Alur, R., Henzinger, T.A.: A really temporal logic. J. ACM **41**(1), 181–203 (1994)
3. Bartocci, E., Corradini, F., Merelli, E., Tesei, L.: Detecting synchronisation of biological oscillators by model checking. Theor. Comput. Sci. **411**(20), 1999–2018 (2010). Hybrid Automata and Oscillatory Behaviour in Biological Systems
4. Basin, D., Klaedtke, F., Müller, S., Pfitzmann, B.: Runtime monitoring of metric first-order temporal properties. In: Hariharan, R., Mukund, M., Vinay, V. (eds.) Proceedings of the 28th IARCS Annual Conference on Foundations of Software Technology and Theoretical Computer Science (FSTTCS 2008) (Dagstuhl, Germany, 2008). Schloss Dagstuhl - Leibniz-Zentrum fuer Informatik, Germany (2008)
5. Brim, L., Dluhoš, P., Šafránek, D., Vejpustek, T.: STL: extending signal temporal logic with signal-value freezing operator. Inf. Comput. **236**, 52–67 (2014)
6. Burattin, A., Maggi, F.M., Sperduti, A.: Conformance checking based on multiperspective declarative process models. Expert. Syst. Appl. **65**, 194–211 (2016)
7. Chomicki, J., Toman, D.: Temporal Logic in Information Systems. In: Chomicki, J., Saake, G. (eds.) Logics for Databases and Information Systems, vol. 436, pp. 31–70. Springer, Heidelberg (1998). https://doi.org/10.1007/978-1-4615-5643-5_3
8. Chomicki, J., Toman, D.: Temporal logic in database query languages. In: Liu, L., Özsu, M.T. (eds.) Encyclopedia of Database Systems, pp. 2987–2991. Springer, Heidelberg (2009). https://doi.org/10.1007/978-0-387-39940-9_402
9. de Leoni, M., van der Aalst, W.: Data-aware process mining: Discovering decisions in processes using alignments. In: Proceedings of the 28th ACM Symposium on Applied Computing (SAC 2013) 18–22 March, Coimbra, Portugal, pp. 113–129 (2013)

10. Demri, S., Lazić, R.: LTL with the freeze quantifier and register automata. ACM Trans. Comput. Logic 10(3), 16:1–16:30 (2009)
11. Fages, F., Rizk, A.: On temporal logic constraint solving for analyzing numerical data time series. Theor. Comput. Sci. 408(1), 55–65 (2008)
12. Feng, S., Lohrey, M., Quaas, K.: Path checking for MTL and TPTL over data words. In: Potapov, I. (ed.) DLT 2015. LNCS, vol. 9168, pp. 326–339. Springer, Cham (2015). https://doi.org/10.1007/978-3-319-21500-6_26
13. Koymans, R.: Specifying real-time properties with metric temporal logic. Real-Time Syst. 2(4), 255–299 (1990)
14. Maggi, F.M., Westergaard, M.: Using timed automata for a priori warnings and planning for timed declarative process models. Int. J. Coop. Inf. Syst. 23(01), 1440003 (2014)
15. Maler, O., Nickovic, D., Pnueli, A.: Checking temporal properties of discrete, timed and continuous behaviors. In: Avron, A., Dershowitz, N., Rabinovich, A. (eds.) Pillars of Computer Science. LNCS, vol. 4800, pp. 475–505. Springer, Heidelberg (2008). https://doi.org/10.1007/978-3-540-78127-1_26
16. Mannhardt, F., Blinde, D.: Analyzing the trajectories of patients with sepsis using process mining. In: RADAR+EMISA 2017, CEUR-WS.org, pp. 72–80 (2017)
17. Mannhardt, F., de Leoni, M., Reijers, H.A., van der Aalst, W.M.P.: Balanced multi-perspective checking of process conformance. Computing 98(4), 407–437 (2016)
18. Pesic, M., Schonenberg, H., van der Aalst, W.: DECLARE: full support for loosely-structured processes. In: Proceedings of the 11th IEEE International Enterprise Distributed Object Computing Conference, p. 287. IEEE Computer Society, Washington, DC (2007)
19. Räim, M., Di Ciccio, C., Maggi, F.M., Mecella, M., Mendling, J.: Log-based understanding of business processes through temporal logic query checking. In: Meersman, R., et al. (eds.) OTM 2014. LNCS, vol. 8841, pp. 75–92. Springer, Heidelberg (2014). https://doi.org/10.1007/978-3-662-45563-0_5
20. Rozinat, A., van der Aalst, W.M.P.: Conformance checking of processes based on monitoring real behavior. Inf. Syst. 33(1), 64–95 (2008)
21. Schönig, S., Di Ciccio, C., Maggi, F.M., Mendling, J.: Discovery of multi-perspective declarative process models. In: Sheng, Q.Z., Stroulia, E., Tata, S., Bhiri, S. (eds.) ICSOC 2016. LNCS, vol. 9936, pp. 87–103. Springer, Cham (2016). https://doi.org/10.1007/978-3-319-46295-0_6
22. van der Aalst, W.M.P., de Beer, H.T., van Dongen, B.F.: Process mining and verification of properties: an approach based on temporal logic. In: Meersman, R., Tari, Z. (eds.) OTM 2005. LNCS, vol. 3760, pp. 130–147. Springer, Heidelberg (2005). https://doi.org/10.1007/11575771_11

# A Report-Driven Approach to Design Multidimensional Models

Antonia Azzini[1]([✉]), Stefania Marrara[1], Andrea Maurino[2], and Amir Topalović[1]

[1] Consorzio per il Trasferimento Tecnologico, C2T, Milan, Italy
{antonia.azzini,stefania.marrara,amir.topalovic}@consorzioc2t.it
[2] Dipartiment of Informatics, Systemistics and Communication,
Universitá degli studi di Milano Bicocca, Milan, Italy
maurino@disco.unimib.it

**Abstract.** Today, large organisations and regulated markets are subject to the control of external audit associations, which require the submission of a huge amount of information in the form of predefined and rigidly structured reports. The compilation of these reports requires the extraction, transformation and integration of data from different heterogeneous operational databases. This task is usually performed by developing a software ad hoc for each report, or by adopting a data warehouse and analysis tools, which are now established technologies. Unfortunately, the data warehousing process is notoriously long and error prone, and is therefore particularly inefficient when the output of the data warehousing is represented by a limited number of reports. This article presents "MMBR", an approach that can generate a multidimensional model from the structure of expected reports as data warehouse output. The approach is able to generate the multidimensional model and populate the data warehouse by defining a knowledge base specific to the domain. Although the use of semantic information in data storage is not new, the novel contribution of our approach is represented by the idea of simplifying the design phase of the data warehouse, making it more efficient, by using an industry-specific knowledge base and a report-based approach.

**Keywords:** Multidimensional design · Knowledge base · Report driven methodology

## 1 Introduction

Business reporting is a strategic but heavy activity defined as "the public reporting of operating and financial data by a business enterprise," [1] or the regular provision of information to support decision-makers within organizations. Reporting is a fundamental part of the business intelligence and knowledge management activity and it is strongly required by audit organizations. Reporting activity can be realized in an ad hoc way by means of specific and complex

© IFIP International Federation for Information Processing 2019
Published by Springer Nature Switzerland AG 2019
P. Ceravolo et al. (Eds.): SIMPDA 2017, LNBIP 340, pp. 105–127, 2019.
https://doi.org/10.1007/978-3-030-11638-5_6

softwares, or by involving typical operations of extracting, transforming, and loading (ETL) procedures in coordination with a data warehouse.

Reports can be distributed in printed form, via email or accessed via a corporate intranet. In sectors as banking, reports are required by both National and European Central Bank organizations on regular basis, and all required reports must comply specific templates provided by the organizations themselves.

In particular, reports for auditing are often very specific, and their structure is usually imposed by the supervising organizations (e.g. European Central Bank, or the rating agency Moodys). The data included in the report are, in most cases, not useful for decision making activities due to the "control" nature of these reports. As a consequence, companies are forced to develop complex systems to compute data that are not useful for their business activities. In this context, it is clear that the need to develop a new approach able to support, in a fast and efficient way, the generation of reports is compelling.

In this scenario we propose to adopt a data warehouse as storage system for data, but we introduce a new approach aimed at designing the multidimensional models on the basis of the structure of the report itself in a (semi-)automatic way, in order to reduce the time needed to produce the report. A data warehouse essentially combines information from several heterogeneous sources into one comprehensive database. By combining all of this information in one place, a company can analyze its data in a more holistic way, ensuring that it has considered all the information available. Data warehousing also makes possible data mining, which is the task of searching for patterns in the data that could disclose hidden knowledge.

At the basis of a data warehouse lies the concept of the *multidimensional (MD) conceptual view of data*. The main characteristic of the multidimensional conceptual view of data is the fact/dimension dichotomy, which represents the data in an n-dimensional space. This representation facilitates the data interpretation and analysis in terms of *facts* (the subjects of analysis and related measures) and *dimensions* that represent the different perspectives from which a certain object can be analyzed.

Even if data warehousing benefits are well recognized by enterprises, it is well known that the warehousing process is time consuming, complex and error prone. Today the increasing reduction of the time-to-market of products forces enterprises to dramatically cut down the time devoted to the design ad the development of MD models, which support the evaluation of the key performance indicators of services and products.

There are different ways to design a data warehouse and many tools are available to help different systems to "upload" their data into a data warehouse for analysis purposes. However, all techniques are based on first extracting data from all the individual sources, by then removing redundancies and finally organizing the data into a format that can be interrogated.

As use case for presenting our approach we propose *securitization*, which is known by the literature as the financial practice of pooling various types of contractual debt such as residential mortgages, commercial mortgages, auto loans or credit card debt obligations (or other non-debt assets which generate receivables) and selling their related cash flows to third party investors as securities [2]. Mortgage-backed securities, which are the case study presented in this paper, are a perfect example of securitization. By combining mortgages into one large pool, the issuer can divide the large pool into smaller parts based on each individual mortgage's inherent risk of default and then sell those smaller pieces to investors.

In this scenario we propose the MMBR (Multidimensional Model By Report) approach, which is able to automatically create the structure of a multi dimensional model (*MD* in the follow) and fill it on the basis of a knowledge base enriched with mapping information that depend on the specific application context. The preprocessing phase of the report (often a quite complex Excel file) is based on a table identification algorithm, which is able to extract the information needed to define the MD structure of the data warehouse. The approach has been tested in the context of financial data with the aim to automatically create the reports required by the Italian National Bank and by the European Central Bank.

The term "by report" refers to the capability of our solution to create a multidimensional model starting from a given report (typically expressed as Microsoft Excel files) that must to be filled with real data. MMBR is also able to generate the relational data structure related to the created MD, and it is also in charge of filling both fact and dimensional tables supported by domain ontologies and by mapping information to the operational sources.

In the literature there are many methodologies for creating MDs starting by requirements, but this is the first attempt to define an approach for creating a MD model starting directly from the structure of the final reports only.

The remaining of the paper is organized as follows: Sect. 2 introduces the state of the art. Section 3 presents the proposed approach, while Sect. 4 describes the knowledge base that is a key element in the MMBR methodology, and the report graph. In Sect. 5, the table identification algorithm is presented, while Sect. 6 describes the creation of the MD model. A real example taken from the financial domain is then reported in Sect. 7. Conclusions and final remarks are reported in Sect. 8.

## 2    Related Work

In the literature several approaches for creating conceptual MD schema from heterogeneous data sources have been presented. According to [3], these approaches can be classified into three broad groups:

- *Supply-driven:* starting from a detailed analysis of the data sources these techniques try to determine the MD concepts. By using this approach, it is possible to waste resources by specifying unnecessary information structures, and by not being able to really involve data warehouse users. See for instance [4–6].

- *Demand-driven:* These approaches focus on determining the MD requirements based on an end-user point of view (as typically performed by other information systems), and mapping them to data sources in a subsequent step (see for example [7,8]).
- *Hybrid approaches:* Some authors (see for example [9–11]) propose to combine the two previously presented approaches in order to harmonize, in the design of the data warehouse, the data sources information with the end-user requirements.

All the methodologies available in literature, however, have the goal to create a MD model as general as possible in order to allow the generation of any report. This assumption requires a lot of effort in both the warehouse conceptualization phase and in the ETL procedure design and development. In several industrial contexts, there is the need to produce a limited number of reports only and, sometimes, with a very strict and well defined structure due to auditing rules or for specific business requirements. In the finance domain, for example, banks are required by central authorities and rating agencies to produce very specific reports related to the securization activities they perform.

In the field of the Semantic Web, Bontcheva and colleague [12] present an approach for the automatic generation of reports from domain ontologies encoded in Semantic Web standards like OWL. The novel aspects of their so-called "MIAKT generator" are in the use of the ontology, mainly the property hierarchy, in order to make it easier to connect a generator to a new domain ontology.

Another interesting approach is presented in [13], where the authors propose a framework for designing a semantic data warehouse. They represent the topic of analysis, measures and dimensions in the requirements. In such an approach they derive the MIO (Multidimensional Integrated Ontologies) along with the knowledge from external ontology sources and domain ontologies. Nebot and colleagues [13] propose an approach in which a Semantic Data Warehouse is considered as a repository of ontologies and other semantically annotated data resources. Then, they propose an ontology-driven framework to design multidimensional analysis models for Semantic Data Warehouses. This framework provides means for building an integrated ontology, called the Multidimensional Integrated Ontology (MIO), including the classes, relationships and instances representing the analysis developed over dimensions and measures.

Romero and colleague [14] introduce a user-centered approach to support the end-user requirements elicitation and the data warehouse multidimensional design tasks. The authors explain how the feedback of a user is needed to filter and shape results obtained from analyzing the sources, and eventually produce the desired conceptual schema. In this scenario, they define the AMDO (Automating Multidimensional Design from Ontologies) method, aimed at discovering the multidimensional knowledge contained in the data sources regardless of the user's requirements.

The implemented process derives the multidimensional schema from a conceptual formalization of the domain, by defining a fully automatic supply-driven approach working at the conceptual level. Differently from the idea implemented in this work, based on the report as starting point, they consider the queries as first. Such an identification comes from the categorization they introduced from a first analysis, that divides different contributions within a so-called demand "driven", "supply-driven" or "hybrid" framework. The first one focuses on determining the end-user multidimensional requirements to produce a multidimensional schema; the second one starts from a detailed analysis of the data sources to determine the multidimensional concepts in a re-engineering process. The latter refers to the approaches that combine the two previous frameworks.

Another interesting work aimed at supporting the multidimensional schema design is given by [15], in which the authors propose an extension of their previous work [16]. They follow a hybrid methodology where the data source and the end-user requirements are conciliated at the early stage of the design process, by deriving only the entities that are of interest for the analysis. The requirements are converted from natural language text into a logical format. The concepts in each requirement are matched to the source ontology and tagged. Then, the multidimensional elements such as fact and dimensions are automatically derived using reasoning.

On the other hand, Benslimane and colleague [17] define a contextual ontology as an explicit specification of a conceptualization, while Barkat [18] proposes a complete and comprehensive methodology to design multi-contextual semantic data warehouses. This contribution is aimed to provide a context meta model (language) that unifies the definitions provided in Database literature. This language is considered as an extension of OWL, which is the standard proposed by the W3C Consortium [19] to define ontologies. It is defined by the authors in order to provide a contextual definition of the used concepts, by offering an externalization of the context from the ontology side.

Pardillo and colleagues [20] present an interesting approach aimed at describing several shortcomings of the current data warehouse design approaches, showing the benefits of using ontologies to overcome them. This work is a starting point for discussing the convenience of using ontologies in the data warehouse design. In particular the authors present a set of situations in which ontologies may help data warehouse designers with respect to some critical aspects. Examples are the requirement analysis phase, where new concepts and techniques meaning should be clarified to be used by stakeholders, or the phase of reconciling requirements and data sources.

As also considered in this approach, it is important to underline that a domain specific ontological knowledge allows to enrich a multidimensional model in aspects that have not been taken into account during the requirement analysis or data-source alignment phases, as well as other aspects, like for example the application of statistic functions in order to aggregate data. Table 1 summarizes the main concepts explained into the literature reported by the above mentioned contributions.

**Table 1.** Related work summary.

| Author | Reference | Principal explained topics | Work description |
|---|---|---|---|
| Bontcheva et al. | [12] | Semantic Web ontologies | Report automatic generation from property hierarchy encoded in Semantic Web (OWL) |
| Nebot et al. | [13] | Semantic Multidimensional | Multidimensional Integrated Onto. definition with Semantic DW as repository |
| Romero et al. | [14] | User-centered AMDO | User-centered app. to user support and automated multidim. design from ontologies |
| Thenmozhi et al. | [15] | Hybrid approach | Matching among req. concepts to the source ontology and tagging process |
| | [16] | Data source end-user req. | Conciliation among sources user req. for interesting entity extraction |
| Benslimane et al. | [17] | Contextual ontology | Contextual ontology definition and explicit specification of a conceptualization |
| Barkat et al. | [18] | Multi-context semantic DW | Context meta model language as OWL extension for semantic DW definition |
| Pardillo et al. | [20] | Ontologies DW design | Use ontologies for DW design to overcome the shortcomings of DW design |

## 3  Description of the Approach and Outline of the Architecture

The MMBR approach main phases are shown in Fig. 1: (1) Table Processing (TP), (2) Row and Column Header Identification and Extraction (RCHIE), (3) Ontology Annotation (OA), (4) Management of Non-Identified labels (MNL), (5) creation of the MD model, (6) ETL Schema Generation (ETL), and, finally, (7) the Report Generation (RG). The input of the TP phase is the template file that has to be filled with the data extracted from an Operational Data Base (ODB). In the TP phase the preprocessing of the template is performed by removing icons and other figures, moreover all terms in the schema are lowered and comment and description fields are removed.

The RCHIE phase is based on the *table identification algorithm* aimed at identifying and extracting the row and column headers in the template. The details of the table identification algorithm are presented in Sect. 5.

The list of terms recognized in the reports by the table identification algorithm is then annotated on the basis of a knowledge base (see Sect. 4). This phase produces two lists; the first one is the list of identified terms annotated w.r.t. the knowledge base, the second one is the list of terms that are not annotated. There are several possible reasons of failure for the annotation activity. The most frequent reason is that a given term may be not included in the knowledge base because it is not relevant to the domain (e.g. "Total"). It is also possible that a term is not annotated because it is a composition of different terms (such as "MortageLoan" or "DelinquentLoan")[1]. Moreover some terms are written in a language different from English (e.g. "garantito" that means guaranteed in

---

[1] The description of these terms is reported in Sect. 7.

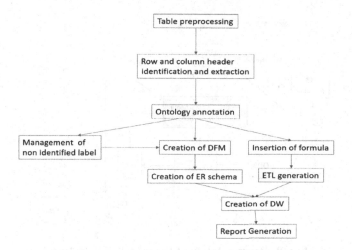

**Fig. 1.** Overall representation of the approach.

Italian). In all these cases, not annotated terms are manually checked and, if relevant, added to the ontology by defining the corresponding rdf:label property. The annotated list of terms is the input for the creation of the dimensional fact model (see Sect. 6). This logical model is translated into a relational star schema. In this phase the relational database is filled with data coming from the ODB. This activity is performed on the basis of the mapping rules included in the knowledge base. This activity is fully described in Sect. 4. Once the data warehouse is filled, the report generation phase is in charge of populating the report template by translating annotation of the report graph into SQL queries executed over the data warehouse. Query results are then inserted in the report template to generate the final output.

The architecture supporting the MMBR approach is represented in Fig. 2. The Annotation Editor is in charge of the first three phases of the MMBR approach, by removing non relevant strings and images from the input file (e.g. logo, comments), and by identifying the terms that are annotated w.r.t. the KB and by creating the report Graph. The Schema builder is the software component aimed at creating the logical relational description of the MD model. The ETL generator is in charge of extracting, on the basis of the Report Graph and the KB, the information necessary to create the extraction-transformation-load data from the ODB to the data warehouse. The Knowledge base manager is in charge of managing and evolving the knowledge base. Any popular tool as, for instance, Protege[2] may be used for the KB creation. The Report generator finally allows to fill the report template by capturing the data from the DW according to the queries build on the base of the annotation included in the Report Graph.

---

[2] https://protege.stanford.edu/.

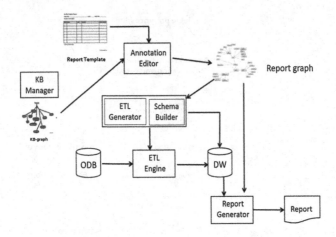

**Fig. 2.** Representation of the overall architecture.

## 4   The MMBR Knowledge Base and the Report Graph

At the core of the proposed approach lies the creation of the knowledge base $KB$, which includes:

- the set of MD concepts and relations (fact, dimensions, measures, attributes);
- the list of terms adopted in the specific application domain (eg. ecommerce, bank securitization,...);
- the Operational DataBase (ODB) schema.

In order to create a knowledge base that could be easily shared in the financial domain we started by using an already existing ontology and only in case of need we created new concepts.

The ontology we used as starting point for creating new MD concepts in the KB is a simplified version of the *data cube vocabulary*[3], i.e., a W3C recommendation for modeling multidimensional data. The top level representation of the defined KB is shown in Fig. 3.

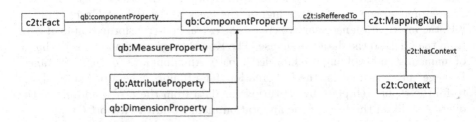

**Fig. 3.** Top level representation of the knowledge base.

---

[3] https://www.w3.org/TR/vocab-data-cube/.

The MD concepts are organized as follows. A fact (the event that is the target of a report, e.g., a sell in a e-commerce domain, a loan in the bank domain) is described by a set of measures and can by analyzed by considering its dimension and descriptive attributes. In the data cube vocabulary dimensions, measures and descriptive attributes are described by the concept *component properties instances*. Dimensions, measures and descriptive attributes are *terms* of the application domain and they are defined by the human (domain) expert trough the Knowledge Base. In fact, the KB annotation specifies if a KB component refers to a fact, a measure or to a dimension. Such elements are then compared with each label extracted from the Excel file in order to define fact, measures and dimensions of the corresponding model. In order to build a KB related to the e-commerce domain, it is possible, for example, to use concepts described in the *good relation* section of the vocabulary[4]. In this scenario instances of *Dimension-Properties* are *gr:ProductOrService, gr:Brand*, while instances of *dq:Measure* are *gr:UnitPriceSpecification, gr:amountOfThisGood*. If no vocabulary is available, a new, ad-hoc, vocabulary has to be defined as first (as also reported in Sect. 7).

The concept *qd:ComponentProperty* can have one or more *rdf:label properties* associated to, that represent the references to the instances of the target concept. For example the dimension *gr:Brand* may be labeled as *"NameOfProduct"* or *"BrandName"*. During the annotation phase, labels are used to associate terms of the report to the application domain concepts.

In order to populate the MD model it is necessary to know how the *qb:componentProperties* are described in the operational DB (ODB). This mapping is described in the KB itself, by means of the *c2t:mappingRule* concept, which associates a *c2t:mappingFormula* related to a given instance of the *qb:ComponentProperties*. The *c2t:mappingFormula* contains a reference to some tables of the ODB and a query predicate over their tuples.

For example, in a bank scenario we can assume that the TLoan table of the ODB contains all information related to loans. A loan with a *fixed rate* (i.e., a loan where the interest rate on the note remains the same through the term of the loan) can be represented in the ODB by the predicate *InterestRate = 1*, while a floating rate can be described by the predicate *InterestRate > 1*. The formula *c2t:mappingFormula* includes the references to the TLoan table and the predicate regarding the *InterestRate*.

The concept *c2t:context* in Fig. 3 assumes value when the reports provided by different audit authorities contain different mapping formulas for the dimension *dq:componentProperties*. For example, a given audit authority may classify a company as "small" if the employee number does not reach 10, while for another authority a company is small if it has less than 15 people employed. In this case we will have two different *c2t:MappingFormula*.

---

[4] http://www.heppnetz.de/projects/goodrelations/.

### 4.1 Report Graph

Starting from such a schema the ontology has been defined by using the Protégé editor [21]. Protégé is a free, open source, ontology editor and a knowledge management system with an user friendly graphic interface. It also includes some classifiers to validate that models are consistent and to infer new information based on the analysis of an ontology [21,22].

The report graph is a rdf representation of the report template, which includes the annotation of each value cell in terms of KB elements.

The top level of the ontology represents the description of the structure of the reports. It includes a set of *qd:observation* elements (i.e., each *cell* of the report), each of them characterized by the two properties (*c2t:hasPosX* and *c2t:hasPosY*) representing their coordinates in the report. An observation element may contain a measure (*c2t:hasValue*) if it contains values from the ODB aggregated by means of an aggregation operator (e.g. *sum, avg...*) as in the traditional data warehouse. Moreover an observation element includes the dimensions, which are used for the analysis. Both dimensions and measure are fully described in the KB. An example of a report graph annotation is as follows:

```
eg:o1 a qb:Observation;
  c2t:hasValue fibo:outstandingPrincipal;
    c2t:isAggregatedBy "Sum";
  c2t:hasPosX "C";
  c2t:hasPosY "9";
    c2t:hasDimensionalProperty ontoLoan:Performing;
    c2t:hasDimensionalProperty ontoLoan:Mortage;
```

In the example, the cell whose coordinates are column "c" and row "9" will contain the sum of *Outstanding Principal* extracted from the loans that are at the same time *Performing* and *with a mortgage guarantee*.

## 5    Table Identification

Reports required by audit organizations are usually structured documents represented by tables. Each table can be divided into different areas or sections, according to their structure. Thus, being able to correctly identify the inner structure of the table is important to find the concepts relevant to the MD models generation. As discussed in the introduction, a multidimensional model represents the data into a n-dimensional space; under this perspective each report can be considered as one of the possible hyperplanes slicing the n-dimensional cube of data. To represent this hyperplane into a bi dimensional table it is necessary to reduce the dimensions. In Fig. 4 the MD is composed by three dimensions (time, nations and type of sold goods) that are "flattened" into a bi-dimensional space by associating the values of type of sold goods (Food and non Food) to

| Time | Thailand | | Japan | | Total |
|------|----------|---------|-------|---------|-------|
|      | Food | NonFood | Food | NonFood |       |
| 2006 | 2400 | 2200 | 11000 | 5000 | 20600 |
| 2007 | 4500 | 3200 | 12000 | 6000 | 25700 |
| 2008 | 5600 | 2900 | 10000 | 5500 | 24000 |
| Total | 12500 | 8300 | 33000 | 16500 | 70300 |

**Fig. 4.** Example of report.

the nation dimension. According to this assumption, rows and columns header may contain dimensions, values of dimensions and measures of the MD.

In our approach, during the RCHIE phase a table was assumed as composed by three types of cell: respectively *textual*, *data* and *schema* ones. Figure 5 shows the general schema. The cell identifiers are represented by the couple $<X,Y>$, as reported in the table shown in Fig. 5.

| <1,1> | abc | abc |          |          |     | <1,n>    |
|-------|-----|-----|----------|----------|-----|----------|
|       |     |     | C_Header | C_Header | ... | C_Header |
| abc   |     |     | C_Header | C_Header | ... | C_Header |
| abc   | R_Header | R_Header | <4,4> |     |     |          |
|       | R_Header | R_Header |       |     | ... |          |
|       | ... | ... |          |          |     |          |
| <m,1> | R_Header | R_Header |       |     |     | <m,n>    |

**Fig. 5.** Example of table used by the table identification algorithm.

The table may contain several types of cells, as defined in the following way:

- **textual-cell:** this cell is not used for table annotation, these cells are shown in grey in Fig. 5, and they may contain simple text.
- **data-cell:** it contains data that are computed on the basis of the MD model. These cells are represented by the white colour in the figure.
- **schema-cell:** it specifies properties over a set of data-cells. It is shown in dark grey in the figure. This cell defines the header $h = <x,y>$ of a set of data cells, by specifying some semantic aspects (i.e., the measure or a value on a dimension).

Rows and columns are identified in order to extract the labels corresponding, respectively, to measures, dimensions, instances of the dimensions, etc. (for instance not relevant information as the $TOTAL$ value shown in Fig. 6). These labels represent the input of the annotation phase, which produces the annotated list of terms as output.

In the literature different table identification algorithms aimed at handling the tables structure have been proposed [23]; in our work the focus is identifying

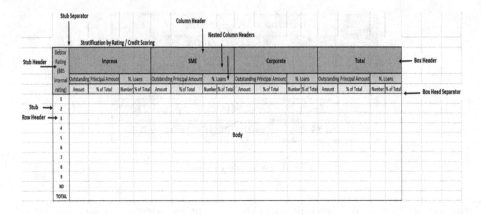

**Fig. 6.** Example of table.

and removing multi spanning cells. An example is reported in Fig. 6, where one of the reports related to securitization is shown. The *Stub Header* cell details information w.r.t. the measures *Loan* and *Outstanding Principal* of different types of companies, as Corporate, SME and "Impresa" (it refers to retail companies in the Italian jargon). Measures, names and instances of dimensions are placed in the *Box Header* and/or the *Stub* areas as headers, and they are used to index the elements located in the *Body* area of the table. The *Stub Header* may also contain a header naming or describing the dimensions located in the stub. The outcomes of the table identification algorithm are shown in Fig. 7 where all data-cells are semantically associated to their row and column headers.

Finally, the RCHIE phase extracts a list of unique terms that are in the column and row headers. These terms are then annotated by means of the knowledge base, by evaluating the labels related to the application domain concepts and the terms extracted from the report table. In this task, the domain expert are forced to take action only for those text strings that do not have the corresponding ontology term (an example is given by the string `performing` and the term `in Bonis`).

# 6   The Dimensional Fact Model

Generally speaking, the Dimensional Fact Model (DFM) [24] is a graphical conceptual model for data warehouse aimed to:

- effectively support the conceptual project,
- define an environment over which intuitively define the queries of a user,
- allow the interaction between the designer and the final user for specific request refinements,
- produce useful and non ambiguous documentation.

| | Impresa | Impresa | Impresa | Impresa | SME | SME | SME | SME | Corporate | Corporate | Corporate | Corporate | Total | Total | Total | Total |
|---|---|---|---|---|---|---|---|---|---|---|---|---|---|---|---|---|
| | Outstanding Principal Amount | Outstanding Principal Amount | N. Loans | N. Loans | Outstanding Principal Amount | Outstanding Principal Amount | N. Loans | N. Loans | Outstanding Principal Amount | Outstanding Principal Amount | N. Loans | N. Loans | Outstanding Principal Amount | Outstanding Principal Amount | N. Loans | N. Loans |
| Debtor Rating (BBS internal rating) | Amount | % of Total | Number | % of Total | Amount | % of Total | Number | % of Total | Amount | % of Total | Number | % of Total | Amount | % of Total | Number | % of Total |
| 1 | posx=3, posy=6 | posx=4, posy=6 | -- | | | | | | | | | | | | | |
| 2 | posx=3, posy=7 | posx=4, posy=7 | -- | | | | | | | | | | | | | |
| 3 | posx=3, posy=8 | posx=4, posy=8 | -- | | | | | | | | | | | | | |
| 4 | posx=3, posy=9 | posx=4, posy=9 | -- | | | | | | | | | | | | | |
| 5 | | | | | | | Body | | | | | | | | | |
| 6 | | -- | -- | | | | | | | | | | | | | |
| 7 | -- | -- | -- | | | | | | | | | | | | | |
| 8 | -- | -- | -- | | | | | | | | | | | | | |
| 9 | -- | -- | -- | | | | | | | | | | | | | |
| NO | -- | -- | -- | | | | | | | | | | | | | |
| TOTAL | -- | -- | -- | | | | | | | | | | | | | |

**Fig. 7.** Example of flattened table.

The conceptual representation deriving from DFM is defined by a set of fact schema. The basic elements modeled by such a schema are the so called fact, measures and dimensions. A fact is useful for the decisional process: it models a set of events coming from the analysis context; it needs to be time evolving. A measure represents a numeric property of a fact, and it describe a quantitative aspect useful for further analysis. Finally, a dimension is a property with a finite domain of a fact and it describes one of the analysis coordinates.

According to the literature [4] a dimensional scheme consists of a set of fact schemes. The components of fact schemes are facts, measures, dimensions and hierarchies. A fact is a focus of interest for the decision-making process; typically, it models an event occurring in the enterprise world (e.g., sales and shipments). Measures are continuously valued (typically numerical) attributes which describe the fact from different points of view; for instance, each sale is measured by its revenue. Dimensions are discrete attributes which determine the minimum granularity adopted to represent facts; typical dimensions for the sale fact are product, store and date. Hierarchies are made up of discrete dimension attributes linked by one-to-one relationships, and determine how facts may be aggregated and selected significantly for the decision-making process. The dimension in which a hierarchy is rooted defines its finest aggregation granularity; the other dimension attributes define progressively coarser granularities. Hierarchies may also include non-dimension attributes. A non-dimension attribute contains additional information about a dimension attribute of the hierarchy, and is connected by a one-to-one relationship (e.g., the address); unlike dimension attributes, it cannot be used for aggregation. At a conceptual level, distinguishing between measures and dimensions is important since it allows the logical design to be more specifically aimed at the efficiency required by data warehousing applications.

## 6.1   Queries Representation

In general, querying an information system means linking different concepts through user defined paths in order to retrieve some data of interest; in particular, for relational databases this is done by formulating a set of joins to connect relation schemes. On the other hand, a substantial amount of queries on DWs are aimed at extracting summary data to fill structured reports to be analysed for decisional or statistical purposes. Thus a typical DW query can be represented by the set of fact instances, at any aggregation level, whose measure values have to be retrieved.

The sets of fact instances can be denoted by writing fact instance expressions. The simple language proposed in the literature [24] is aimed at defining, with reference to a dimensional scheme, the queries forming the expected workload for the DW, to be used for logical design; thus, it focuses on which data must be retrieved and at which level they must be consolidated.

A fact instance expression has the general form:

```
<fact instance expression> ::= <fact name>
             (<pattern clause> ; <selection clause>)
<pattern clause> ::= comma-list of <pattern elements>
<pattern elements> ::= <dimension name> |
                       <dimension name>.<attribute name>
<selection clause> ::= comma-list of <predicate>
```

The pattern clause describes a pattern. The selection clause contains a set of Boolean predicates which may either select a subset of the aggregated fact instances or affect the way fact instances are aggregated. If an attribute involved either in a pattern clause or in a selection clause is not a dimension, it should be referenced by prefixing its dimension name.

## 6.2   MD Creation and Population on MMBR Approach

The Dimensional Fact Model (DFM) [4] approach has been used in our solution to describe the MD model. The list of annotated terms and the KB are the only two elements necessary to design and populate the MD. Each annotated term of the list is enriched by its type or subclass in order to understand if it is a measure, a dimension or an instance of a dimension. This can be realized by means of a set of SPARQL[5] queries over the KB (an example of query is shown in Sect. 7) generated by the Schema builder component. With this information it is possible to create the DFM and the corresponding logical relational schema by means of the original methodology proposed in [4]. The relational schema is then populated according to the mapping information defined in the knowledge base.

All dimensional tables are populated with the instances defined in the KB, while the fact table is defined in a two steps procedure. In the first step all

---

[5] https://www.w3.org/TR/rdf-sparql-query/.

instances of the facts (e.g. sell or loan) are selected from the ODB by taking into account only the measures available in the annotated list. The second step is in charge of connecting the fact table with the dimensional tables. An Update query is executed to associate each instance of the fact table with the instances of the dimensions tables. Even in this case the KB plays a strategic role since it allows to extract the mapping formula at the basis of the SPARQL queries (see Sect. 7).

# 7  Case Study

After a brief introduction over the ontology defined in this work, an example of two financial reports that have to be filled is reported, together with an example of the mapping rules and the defined sparql queries. The implemented software prototype, supporting the MMBR approach, and a brief discussion about the methodology are finally presented.

## 7.1  The Considered Reports

The scenario motivating the definition of a report driven approach for the design of multidimensional models is related to the financial domain. In particular, the reporting activity of securitization was analyzed.

Applying the MMBR approach in this context, the first activity to be faced is the generation of the domain KB and vocabulary. The literature proposes two different vocabularies that partially describe the loan domain: FIBO[6] and Schema.org[7]. FIBO, a Financial Industry Business Ontology, contains the loan terms definitions without any further specification. Schema.org does not contain a full exhaustive specification of the securitization domain, but it includes the LoanOrCredit concepts[8] only. The KB defined in this work to describe the securitization domain is the ontology *OntoLoan*. During the KB definition, domain experts are in charge to define the main terms and concepts. The OntoLoan ontology is not freely available, since it is covered by the company's intellectual properties. However, the top level of OntoLoan is shown in Fig. 8, while Fig. 9 shows an example of securization report. Note that all private data related to the bank owning the report are removed for privacy issues, while the values for different kinds of loans are reported.

The term *Performing Loans* refers to those loans that have no overdue interest payments, or with unpaid installments due, even if under the maximum number of delay days outstanding (which changes according to the securitization contract terms). *Delinquent Loan* refers to the loans close to default, i.e., to unpaid installments due to a delay in payments close to the limit on the delay of days overdue. *Defaulted Loans* refers then to loans with significant delays in payments.

---

[6] https://www.edmcouncil.org/financialbusiness.
[7] https://schema.org.
[8] https://schema.org/LoanOrCredit.

Any kind of loan is further divided according to other features, generating the definition of *Mortgage Loan, Guaranteed Loan,* i.e. loans insured not by mortgages but by other guarantees (e.g., pledges), and, finally, *Unguaranteed Loan,* i.e. not insured.

The first phase of the MMBR approach removes text fields that do not carry relevant information from the report. An example of removed test is the string "A. PORTFOLIO OUTSTANDING BALANCE". The annotation tool removes the cell spanning from the table of Fig. 9, creating the table structure shown in Fig. 10. The data-cell in position <3, 3> represents the aggregation of the values of *Outstanding Principal* of loans that are both performing and able to pay off the loan even in case of default of the borrower. The value in the cell with position < 3, 4 > represents the aggregation of the *Outstanding Principal* of loans that are both performing and guaranteed. The mapping rule MR1 is described as follows:

```
MR1   :mappingRule1 rdf:type :MappingRule ;
  :hasContext :context1;
              :hasTargetDimension :defaulted ;
              :hasMappingFormula "rating_34=10 and
              rating5 between 1 and 7"^^xsd:string
```

The rule MR1 indicates that for the context *context1* the *defaulted* value (instance of performance category) is associated to the loans having a *rating34* equals to 10 and a *rating5* between 1 and 7.

## 7.2  SPARQL Queries Definition

With this first activity the following list of terms related to the domain is extracted *loan:Performing, loan:Mortgage, loan:Guaranteed, loan:Unguaranteed, loan:Delinquent, loan:Defaulted, loan:DelinquentInstalments, loan:Outstanding Principal, loan:AccruedInterest, loan:PrincipalInstalment, loan:Interest Instalment.* For each element of the list, MMBR retrieves from the KB the name of the dimensions or measures related to it, by means of *SPARQL* queries.

An example of query is the following.

```
SELECT distinct  ?x, ?p
WHERE {
loan:Guarantee rdf:type ?x.
?x rdfs:subClassOf ?p
}
```

The example query is able to recognize, as shown in Fig. 8, that the *loan:Guarantee* element is member of an entity named *Guarantee_Category*, which is a subclass of *qb:DimensionProperty.* Figure 8 also shows the query properties. After the identification of the measures and the dimensions, the DFM is designed as shown in Fig. 11, according to Literature (see [25]) for the schema definition.

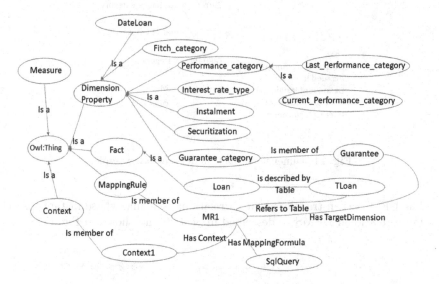

**Fig. 8.** The top level representation of OntoLoan.

|  |  | Outstanding Principal | Accrued Interest | Delinquent instalments | | | Total |
|---|---|---|---|---|---|---|---|
|  |  |  |  | Principal inst. | Interest inst. | Total |  |
| Performing Loans | Mortgage Loans |  |  |  |  |  |  |
|  | Guaranteed Loans |  |  |  |  |  |  |
|  | Unguaranteed Loans |  |  |  |  |  |  |
|  | Total Unsecured |  |  |  |  |  |  |
| Delinquent Loans | Mortgage Loans |  |  |  |  |  |  |
|  | Guaranteed Loans |  |  |  |  |  |  |
|  | Unguaranteed Loans |  |  |  |  |  |  |
|  | Total Unsecured |  |  |  |  |  |  |
| Defaulted Loans | Mortgage Loans |  |  |  |  |  |  |
|  | Guaranteed Loans |  |  |  |  |  |  |
|  | Unguaranteed Loans |  |  |  |  |  |  |
|  | Total Unsecured |  |  |  |  |  |  |
|  | TOTAL |  |  |  |  |  |  |

**Fig. 9.** An example of report template.

|  |  | Outstanding Principal | Accrued Interest | Delinquent instalments | Delinquent instalments | Delinquent instalments | Total |
|---|---|---|---|---|---|---|---|
|  |  |  |  | Principal inst. | Interest inst. | Total |  |
| Performing Loans | Mortgage Loans | pos=2,pos.p.3 |  |  |  |  |  |
| Performing Loans | Guaranteed Loans | pos=3,pos.p.4 |  |  |  |  |  |
| Performing Loans | Unguaranteed Loans | pos=3,pos.p.5 |  |  |  |  |  |
| Performing Loans | Total Unsecured |  |  |  |  |  |  |
| Delinquent Loans | Mortgage Loans |  |  |  |  |  |  |
| Delinquent Loans | Guaranteed Loans |  |  |  |  |  |  |
| Delinquent Loans | Unguaranteed Loans |  |  |  |  |  |  |
| Delinquent Loans | Total Unsecured |  |  |  |  |  |  |
| Defaulted Loans | Mortgage Loans |  |  |  |  |  |  |
| Defaulted Loans | Guaranteed Loans |  |  |  |  |  |  |
| Defaulted Loans | Unguaranteed Loans |  |  |  |  |  |  |
| Defaulted Loans | Total Unsecured |  |  |  |  |  |  |
|  | TOTAL |  |  |  |  |  |  |

**Fig. 10.** An example of flattened report.

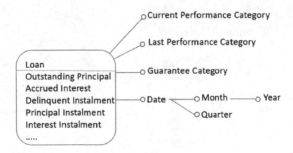

**Fig. 11.** Dimensional fact model schema example.

The DFM is then translated into a relational schema, whose instance is created in a relational DBMS as described in Sect. 6, and shown in Fig. 12.

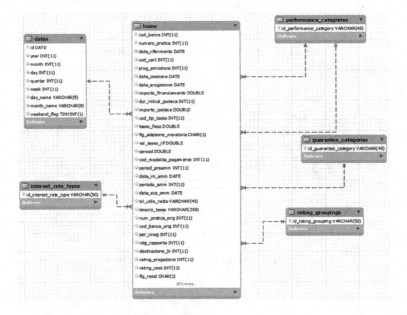

**Fig. 12.** Data warehouse schema.

In order to update the fact table, it is possible to retrieve the mapping formula in the KB, by means of a SPARQL query. For example to update the *guaranteed loan*, first we retrieve from the KB the corresponding mapping formula by using the following query:

```
SELECT ?table, ?rule
WHERE {
?s rdf:type loan:MappingRule.
?s loan:hasContext loan:context1.
?s  loan:hasTargetDimension loan:Guarantee.
?s  loan:refersToTable ?table.
    ?s  loan:hasMappingFormula ?rule.
}
```

The result is the following predicate:

```
TLoan
VAL_IPOTECA = 0 and (flag_garanzia_confidi='Y' or
(importo_pegno + importo_garan_pers) > 0)"?^^string
```

The corresponding update query using IBM DB2 SQL is:

```
UPDATE Fact
SET id_Guarantee_category=
(SELECT \Guarantee"?
    FROM fact join odb.TLoan
    WHERE fact.id=obd.TLoan.id and
    VAL_IPOTECA = 0 and
    (flag_garanzia_confidi='Y' or (importo_pegno + importo_garan_pers) > 0)
)
```

The last phase is related to the generation of the report. Let us assume to generate a report where the cell corresponding to the coordinates F and 22 contains the sum of interest installments of all defaulted and guarantee loans. In the report graph the cell F22 is annotated as follows:

```
#Cell F22
eg:o36 a qb:Observation;
    c2t:hasDimensionalProperty ontoLoan:Defaulted;
    c2t:hasDimensionalProperty ontoLoan:Unguaranteed;
    c2t:hasValue ontoloan:InterestInstalment;
    c2t:isAggregatedBy "Sum"
    c2t:hasPosY "22";
    c2t:hasPosX "F";
```

The ReportGenerator module creates the aggregate SQL query able to compile the cell. First of all, it retrieves the correct aggregation operator to be used, i.e. here it is a *sum*, then by querying the rdf fragment of the report graph it discovers that the values to be aggregated belong to the attribute *InterestInstalment*. To create the FROM statement of the query the ReportGenerator interrogates the ontology finding that both *ontoLoan:Defaulted* and *ontoLoan:Unguaranteed* concepts are instances of *ontoLoan:DimensionalProperty* and that both attributes are included in the Loans table. Thus, the FROM condition includes the Loans table only. The WHERE condition is composed creating a conjunctive predicates of $current\_performance\_category = Defaulted$ and $guarantee\_category = Unguaranteed$. The final SQL query for computing the value of the Cell F22 is the following:

```
SELECT sum(InterestInstalment)
FROM Loans
WHERE current_performance_category="Defaulted" AND
  guarantee_category="Unguaranteed"
```

The report generated is show in Fig. 13.

| | | Outstanding Principal | Accrued Interest | Delinquent instalments | | Total | Total |
| | | | | Principal inst. | Interest inst. | Total | |
| | | a | b | c | d | e = c + d | f = a + b + e |
| 1 | Performing Loans | 784.388.232,49 | 2.117.994,43 | 1.186.477,11 | 271.017,76 | 1.457.494,87 | 787.963.721,79 |
| | Mortgage Loans | 436.453.328,80 | 1.490.251,88 | 487.789,26 | 143.702,04 | 631.491,30 | 438.575.071,98 |
| | Guaranteed Loans | 256.920.537,40 | 406.690,11 | 574.366,19 | 95.721,59 | 670.087,78 | 257.997.315,29 |
| | Unguaranteed Loans | 91.014.366,29 | 221.052,44 | 124.321,66 | 31.594,13 | 155.915,79 | 91.391.334,52 |
| | Total Unsecured | 347.934.903,69 | 627.742,55 | 698.687,85 | 127.315,72 | 826.003,57 | 349.388.649,81 |
| 2 | Delinquent Loans | 10.015.697,28 | 55.716,45 | 370.642,23 | 101.403,52 | 472.045,75 | 10.543.459,48 |
| | Mortgage Loans | 5.677.758,73 | 47.854,50 | 26.801,82 | 74.087,28 | 100.889,10 | 5.826.512,43 |
| | Guaranteed Loans | 3.463.351,91 | 8.496,19 | 236.242,02 | 21.191,46 | 257.433,48 | 3.727.281,58 |
| | Unguaranteed Loans | 874.586,64 | 1.355,66 | 107.598,39 | 6.124,78 | 113.723,17 | 989.665,47 |
| | Total Unsecured | 4.337.938,55 | 7.851,85 | 343.840,41 | 27.316,24 | 371.156,65 | 4.716.047,05 |
| 3 | Collateral Portfolio (1+2) | 794.403.929,77 | 2.173.710,88 | 1.557.119,34 | 372.421,28 | 1.929.540,62 | 798.507.181,27 |
| 4 | Defaulted Loans | 9.468.456,24 | 53.897,77 | 606.766,09 | 141.486,02 | 748.252,11 | 10.270.606,12 |
| | Mortgage Loans | 5.709.138,55 | 40.931,49 | 180.150,10 | 64.062,84 | 244.212,94 | 5.994.282,98 |
| | Guaranteed Loans | 3.236.549,82 | 9.426,96 | 313.545,45 | 67.325,72 | 380.871,17 | 3.626.847,95 |
| | Unguaranteed Loans | 522.767,87 | 3.539,32 | 113.070,54 | 10.097,46 | 123.168,00 | 649.475,19 |
| | Total Unsecured | 3.759.317,69 | 12.966,28 | 426.615,99 | 77.423,18 | 504.039,17 | 4.276.323,14 |
| | TOTAL ( 3+4) | 803.872.386,01 | 2.227.608,65 | 2.163.885,43 | 513.907,30 | 2.677.792,73 | 808.777.787,39 |

**Fig. 13.** Report generated.

### 7.3   Discussion

The MMBR methodology and related techniques supporting the creation of multidimensional model able to produce a given (set of) report(s). The term "by report" refers to the capability of our solution to create a multidimensional model starting from a given report (typically expressed as Microsoft Excel file) that must to be filled with real data. MMBR is also able to generate the relational data structure related to the created a MD model and it is also in charge of filling both fact and dimensional tables thanks to the use of domain ontologies enriched with mapping information to the operational sources. According to the literature, this is the first attempt to define a methodological approach for creating MD starting directly from the reports only. The methodology starts with the acquisition of the excel file and, thanks to an table identification algorithm, then it extracts rows and headers representing domain concept from the report, and the extracted terms are annotated by using a domain ontology enriched with md concepts. The ontological terms are finally used to design the MD.

One of the most important elements in our methodology is the use of a domain ontology, applied in order to annotate terms available in the report. Such ontological terms are used to identify fact, dimensions and instances of dimension that allow the creation and population of the MD model, by generating the Dimensional Fact model and the Entity Relational schema.

A prototype supporting the proposed methodology has been developed in the experimental session. The report graph is created by using *Protegé*, which is adopted to support the RCHIE phase in a semiautomated way. The phases of the methodology involved into the creation and population of the data warehouse are supported by a custom tool we named "CreDaW", (Create a Data Warehouse).

The tool, as described in Sect. 6, creates the DW schema by querying the Report graph. The relational tables implementing the DW schema are populated by querying the KB as reported in Sect. 7. The prototype is developed and tested on an Intel I7-6700 personal computer with 3.4 GHz, 16 GB ram and 1 TB hdd and it is able to create and populate a DW in two different relational database management systems, MySQL and IBM DB2.

The data warehouse population phase requires around 13 s for loading more than 400.000 loans.

The last phase (the "Report generation") of the methodology is totally automated by the tool *ReGe*. The tool is able to read the report graph and, by using the Apache POI library (https://poi.apache.org/), the report template. For each observation in the report graph a SQL query is created and executed; the result fills the corresponding cell of the report template. A report generation requires less than 2 s.

## 8    Conclusions and Future Work

This work presents a "Multidimensional Model By Report" (MMBR) approach supporting the creation of multidimensional models able to produce a given (set of) report(s). The term "by report" refers to the ability to create a multidimensional (MD) model starting from a given report (typically expressed as Microsoft Excel file) that has to be filled with data extracted from a set of heterogeneous sources.

Important contributions refer to the automatic generation of the relational data structure correlated to the MD models generated by the approach, and to the ability to fill both fact and dimensional tables on the basis of domain ontologies enriched with mapping information related to the data sources.

There may be several future directions of research. The first one is related to the definition of an approach for the automatic computation of aggregates of data according to the topological position of the cells that contain them, by taking into account row and column headers.

Another interesting research activity will study how to enrich the table identification algorithm. The aim is to allow the management of a larger (w.r.t., the actual algorithm) number of types of report, improving the efficiency of the presented approach.

# References

1. Lymer, A., Debreceny, R., Gray, G.: Business Reporting on the Internet (1999)
2. Simkovic, M.: Competition and crisis in mortgage securitization. Indiana Law J. **88**, 213 (2013)
3. Winter, R., Strauch, B.: A method for demand-driven information requirements analysis in data warehousing projects. In: 36th Hawaii International Conference on System Sciences (HICSS-36 2003), CD-ROM/Abstracts Proceedings, HI, USA, 6–9 January 2003, p. 231. IEEE Computer Society, Big Island (2003)
4. Golfarelli, M., Maio, D., Rizzi, S.: The dimensional fact model: a conceptual model for data warehouses. Int. J. Coop. Inf. Syst. **7**(2–3), 215–247 (1998)
5. Golfarelli, M., Graziani, S., Rizzi, S.: Starry vault: automating multidimensional modeling from data vaults. In: Pokorný, J., Ivanović, M., Thalheim, B., Šaloun, P. (eds.) ADBIS 2016. LNCS, vol. 9809, pp. 137–151. Springer, Cham (2016). https://doi.org/10.1007/978-3-319-44039-2_10
6. Blanco, C., de Guzmán, I.G.R., Fernández-Medina, E., Trujillo, J.: An architecture for automatically developing secure OLAP applications from models. Inf. Softw. Technol. **59**, 1–16 (2015)
7. Jovanovic, P., Romero, O., Simitsis, A., Abelló, A., Mayorova, D.: A requirement-driven approach to the design and evolution of data warehouses. Inf. Syst. **44**, 94–119 (2014)
8. Prat, N., Akoka, J., Comyn-Wattiau, I.: A UML-based data warehouse design method. Decis. Support Syst. **42**(3), 1449–1473 (2006)
9. Nabli, A., Feki, J., Gargouri, F.: Automatic construction of multidimensional schema from OLAP requirements. In: 2005 ACS/IEEE International Conference on Computer Systems and Applications (AICCSA 2005), 3–6 January 2005, Egypt, p. 28. IEEE Computer Society, Cairo (2005)
10. Giorgini, P., Rizzi, S., Garzetti, M.: Grand: a goal-oriented approach to requirement analysis in data warehouses. Decis. Support Syst. **45**(1), 4–21 (2008)
11. Blanco, C., de Guzmán, I.G.R., Fernández-Medina, E., Trujillo, J.: An MDA approach for developing secure OLAP applications: metamodels and transformations. Comput. Sci. Inf. Syst. **12**(2), 541–565 (2015)
12. Bontcheva, K., Wilks, Y.: Automatic report generation from ontologies: the MIAKT approach. In: Meziane, F., Métais, E. (eds.) NLDB 2004. LNCS, vol. 3136, pp. 324–335. Springer, Heidelberg (2004). https://doi.org/10.1007/978-3-540-27779-8_28
13. Nebot, V., Berlanga, R., Pérez, J., Aramburu, M., Pedersen, T.: Multidimensional integrated ontologies: a framework for designing semantic data warehouses. J. Data Semant. XII I, 1–36 (2009)
14. Romero, O., Abelló, A.: A framework for multidimensional design of data warehouses from ontologies. Data Knowl. Eng. **69**(11), 1138–1157 (2010)
15. Thenmozhi, M., Vivekanandan, K.: An ontology based hybrid approach to derive multidimensional schema for data warehouse. Int. J. Comput. Appl. **54**(8), 36–42 (2012)
16. Thenmozhi, M., Vivekanandan, K.: A framework to derive multidimensional schema for data warehouse using ontology. In: Proceedings of National Conference on Internet and WebSevice Computing, NCIWSC (2012)
17. Benslimane, D., Arara, A., Falquet, G., Maamar, Z., Thiran, P., Gargouri, F.: Contextual ontologies. In: Yakhno, T., Neuhold, E.J. (eds.) ADVIS 2006. LNCS, vol. 4243, pp. 168–176. Springer, Heidelberg (2006). https://doi.org/10.1007/11890393_18

18. Barkat, O., Khouri, S., Bellatreche, L., Boustia, N.: Bridging context and data warehouses through ontologies. In: Proceedings of the Symposium on Applied Computing, pp. 336–341. ACM (2017)
19. W3C: W3C Standard Consortium. http://www.w3.org
20. Pardillo, J., Mazón, J.N.: Using ontologies for the design of data warehouses. Int. J. Database Manag. Syst. (IJDMS) **3**(2), 73–87 (2011)
21. Protégé: Protégé Ontology Editor. https://protege.stanford.edu/
22. Nadeau, D., Sekine, S.: A survey of named entity recognition and classification. Lingvisticae Investigationes **30**(1), 3–26 (2007)
23. Zanibbi, R., Blostein, D., Cordy, J.R.: A survey of table recognition. Doc. Anal. Recogn. Models Obs. Transform. Infer. **7**(1), 1–16 (2004)
24. Golfarelli, M., Maio, D., Rizzi, S.: The dimensional fact model: a conceptual model for data warehouses. Int. J. Coop. Inf. Syst. **7**(02n03), 215–247 (1998)
25. Sugumaran, V., Storey, V.C.: Ontologies for conceptual modeling: their creation, use, and management. Data Knowl. Eng. **42**(3), 251–271 (2002)

# Author Index

Printed in the United States
By Bookmasters